ENVISIONING FUTURES

FOR THE SACRAMENTO–SAN JOAQUIN DELTA

JAY LUND | ELLEN HANAK

WILLIAM FLEENOR | RICHARD HOWITT

JEFFREY MOUNT | PETER MOYLE

2007

Library of Congress Cataloging-in-Publication Data

Envisioning futures for the Sacramento–San Joaquin Delta / Jay Lund ... [et al.].

p. cm.

Includes bibliographical references.

ISBN: 978-1-58213-126-9

1. Delta Region (Calif.) 2. Water-supply—California—Delta Region—
Management. 3. Natural resources—California—Delta Region. 4. Flood
control—California—Delta Region. 5. Levees—California—Delta Region.
6. Environmental management—California—Delta Region. I. Lund, Jay R.

TD225.D29E58 2007

333.91'62097945—dc22

2007000040

Cover photo by Roy Tennant.

Foreword

It is telling that Hurricane Katrina would send a warning sign to California—in part because California is generally regarded as a hot bed of natural disasters. Fires, floods, and earthquakes rake the state with a frustrating regularity. Yet this is exactly what happened: Devastation in the South alerted those of us in the West to yet another potential disaster. As Californians turned outward to meet the needs of former residents of the Gulf region, especially those unfortunate enough to have lived in the lower-lying neighborhoods of New Orleans, state policymakers turned inward and realized that the Sacramento Delta held the same loss potential from a major earthquake as New Orleans had experienced from a hurricane.

Ellen Hanak, research fellow and director of the Economy Program at PPIC, and a team of experts from the University of California, Davis, decided to explain the vulnerability of the Sacramento–San Joaquin Delta and to lay out a series of options for addressing current and likely future problems. This report, *Envisioning Futures for the Sacramento–San Joaquin Delta,* describes why the Delta matters to Californians and why the region is currently in a state of crisis—from threatened freshwater supplies for the whole state to the potential extinction of numerous fish species. After reviewing years of policy studies on the Delta, as well as delving into the most updated ecological information, the authors conclude that the future requires a "vision of a variable Delta, as opposed to the commonly held vision of a static Delta." The strategy of rigorously preserving a freshwater Delta has been risky and expensive. Instead, the authors present a case for a future approach that "yields the best outcomes overall, accompanied by strategies to reasonably compensate those who lose Delta services."

Nine alternatives are presented across three objectives—maintaining high levels of fresh water, allowing the Delta to fluctuate between high and low levels of salinity, and moving toward a Delta that provides high levels of fresh water as needed. The authors carry out an initial summary evaluation of all nine alternatives and provide a rationale for their assessment of each one. The report does not endorse any single "best" solution among these alternatives. As the authors note, a closer look at the details will be required before the best strategy can be decided on.

However, they suggest that a hybrid solution, relying on some combination of key elements, may provide the most promising path forward.

In this spirit, the report offers a number of new ideas for managing the Delta and presents a set of central themes for ways to think about the future of the region. The most striking of these themes is that business as usual is unsustainable for current stakeholders. The combined effects of continued land subsidence (that is, sinking land elevations), sea level rise, increasing seismic risk, and worsening winter floods make continued reliance on weak Delta levees imprudent and unworkable over the long term. In very strong language, the authors conclude that significant political decisions will be needed to make major changes in the Delta. Incremental, consensus-based solutions are unlikely to prevent a major ecological and economic catastrophe of statewide significance.

The report concludes with recommendations for several actions—some related to the use of technical and scientific knowledge and others to the design of governance and finance policies. Most important, the authors identify a number of urgent items for debate and policy action. With a substantial base of empirical evidence and a considered assessment of the options, the report is not alarmist—but it does make a strong case that California's future water supply is in serious jeopardy unless the problems of the Sacramento–San Joaquin Delta are dealt with in a thoughtful and timely fashion.

David W. Lyon
President and CEO
Public Policy Institute of California

Summary

"One gains nothing . . . by starting out with the question, 'What is acceptable?' And in the process of answering it, one gives away the important things, as a rule, and loses any chance to come up with an effective, let alone with the right, answer."

Peter F. Drucker (1967), The Effective Executive

California's Delta Crisis

The Sacramento–San Joaquin Delta is the hub of California's water system, home to a unique ecosystem and to a diverse recreational and agricultural economy. Management strategies for the Delta that satisfy these often competing interests have been discussed and debated for almost 100 years, at times leading to acrimonious divisions between Northern and Southern California, environmental and economic interests, and agricultural and urban water users. Recently, the Delta has again taken center stage in debates on California water policy, with broad implications for statewide environmental, land use, and flood control policies. The Delta is widely perceived to be in crisis in several ways.

One dimension of the crisis is the health of the Delta's 1,100 miles of levees, on which both Delta land use and water supply systems depend. The devastating effects of Hurricane Katrina on New Orleans' levees galvanized public attention on the fragility of the Delta's levee system, where close calls occur with some frequency and where a major levee break occurred in June 2004. Continued sinking of Delta islands, sea level rise, and likely increases in the severity of flooding make the Delta's fragile levee network increasingly vulnerable to failure from earthquakes, floods, and other causes.

Long-term increases in these risk factors make the current reliance on Delta levees appear imprudent and unsustainable. Over the next 50 years, there is a two-thirds chance of a catastrophic levee failure in the Delta, leading to multiple island floodings and the intrusion of seawater. For one such scenario, the Department of Water Resources estimates that a large earthquake near the Delta would cause major interruptions in water supplies for Southern California, the San Joaquin Valley, and the Bay Area, as well as disruptions of power, road, and shipping lines, costing the state's economy as much as $40 billion. Such failures also would create major environmental disruptions and local flooding risks.

A second aspect of the crisis is the health of Delta fish species. In the fall of 2004, routine fish surveys registered sharp declines in the numbers of several open-water (pelagic) species, including the delta smelt, already listed as threatened under the federal and state Endangered Species Acts. Subsequent surveys have confirmed the trend, raising concerns that the smelt—sometimes seen as an indicator of ecosystem health in the Delta— risks extinction if a solution is not found quickly.

The third dimension of the crisis is institutional. The framework known as CALFED—a stakeholder-driven process established in the mid-1990s to mediate conflict and to "fix" the problems of the Delta—is facing a crisis of confidence. Although the levee and ecosystem problems noted above are partly to blame, CALFED has also been criticized for failing to elicit anticipated funding commitments. As the CALFED truce erodes, lawsuits are beginning to fill the gaps left by a lack of consensus on management strategies and options. Some of these conflicts reflect a renewal of old battle lines, pitting water exporters against environmental interests and those who use water within the Delta. But new battle lines have also emerged over the urbanization of Delta farmlands and the issue of levee stability. The pressures to develop the Delta's flat, low-lying lands are great, given their location near transportation corridors and several major metropolitan areas. Yet many concerns are being raised about the consequences for flood risk, ecosystem health, and water quality. Moreover, the prospect of levee failure raises concerns about the potentially great financial liabilities facing California's taxpayers, given the state's role in managing the Delta and its many miles of levees.

Responding to the Crisis

Recognition of the crisis in the Delta has led to appeals to pursue a number of very different management strategies. The collapse of key Delta fish populations has prompted some environmentalists to call for cutbacks in water exports. At the same time, two main proposals have surfaced for dealing with levee instability: massive investments in the levee system (creating, in a sense, the "Fortress Delta" we discuss below) or construction of a peripheral canal at the Delta's edge, to protect urban and agricultural interests from what many now view as the unacceptable risks of continued reliance on direct Delta exports. The resurgence of a peripheral canal proposal is significant, because it is a solution that has deeply divided Californians in the past.

As an immediate response to concerns over the health of the levee system, the state significantly increased the budget for levee repairs in 2006, and two bond measures passed in November 2006 allocate additional funds for flood control in the Delta. But there is as yet no broader plan for responding to the crisis in the Delta, including how the bond funds should be spent. Such a plan may emerge from several efforts now under way. Two technical studies are examining the causes of the pelagic organism decline and the risks to the levee system. Two policy-driven efforts are charged with looking at long-term management options. The Delta Vision effort, launched by the governor in the fall of 2006, is to develop a strategic plan for sustainable use of the Delta, in conjunction with a broad range of stakeholders and an independent Blue Ribbon Task Force. During 2007, the CALFED program must also propose alternative management strategies to meet its water and environmental goals for the Delta.

We hope that this study enriches both policy and technical discussions of the Delta's future. Our aim is to begin a serious, scientific search for and comparison of potential long-term solutions for the coming decades. We purposely take a broader view of the options than those commonly under discussion in stakeholder circles—namely, the Fortress Delta, the peripheral canal, and the maintenance of the current levee-centric strategy with lower water export volumes.

The task at hand is urgent, and the stakes in the Delta are high. If California fails to develop a viable solution and act on it soon, we risk the loss of native species and important ecosystem services—and face significant economic disruptions. Yet there is also a risk that the political process will prematurely close off the consideration of options that could help California make the most of the Delta while protecting its unique ecosystem and species.

New Thinking About Solutions for the Delta Ecosystem

For the past 70 years, the state's policy has been to maintain the Delta as a freshwater system through a program of water flow regulation, supported by maintenance of agricultural levees. This strategy improved water quality for Delta agriculture and water exports and was assumed to protect both native and desirable alien species (particularly striped bass). But most such species have not done well under this policy. Native species have declined considerably, and some—including the delta smelt—continue to decline, even to the verge of extinction. Although recent work suggests that export pumping is having a negative effect on several key Delta species, more freshwater inflows or reduced exports alone are unlikely to save these species because the highly altered nature of the aquatic ecosystem is part of the problem.

Before the Delta was drained, diked, and settled by Europeans, it was subject to significant seasonal and interannual fluctuations in freshwater inflows, which worked in concert with large tidal ranges. Some parts of the northern, eastern, and southern Delta were largely fresh at all times. However, the western Delta was seasonally brackish and the central Delta was brackish in the dry seasons of dry years. This was the flow and water quality regime to which many native Delta species are adapted. The invasion of numerous alien species, both as deliberate introductions and as by-products of human activities, has created many problems. Many of these invasive species are better adapted than the natives to the highly altered environment that the Delta has become.

To address the problems of the Delta's native species, a fundamental change in policy is needed. A Delta that is heterogeneous and variable across space and time is more likely to support native species than is a homogeneously fresh or brackish Delta. Accepting the vision of a variable Delta, as opposed to the commonly held vision of a static Delta, will allow for more sustainable and innovative management. This is a legal and political necessity as much as it is an ecological one. Many aspects of Delta water and land management, from export operations to levee maintenance, are significantly affected by a number of federal and state environmental laws. These laws form a significant constraint on any future management strategy of the Delta.

Facing the Tradeoffs

A comprehensive solution for the Delta also needs to take into account goals for the human use of Delta resources—including land use and water supply and quality. But a change in thinking is necessary, particularly in terms of the ability to satisfy all goals simultaneously. The approach adopted by CALFED in the mid-1990s was that "everyone would get better together," and it was assumed that this could be achieved by managing the Delta as a single unit, simultaneously achieving improvements in habitat, levees, water quality, and water supply reliability. Going forward, Californians will need to recognize that the Delta cannot be all things to all people. Tradeoffs are inevitable. The challenge will be to pursue an approach that yields the best outcomes overall, accompanied by strategies to reasonably compensate those who lose Delta services.

Some Alternatives

With this in mind, we consider nine alternative approaches to a comprehensive solution for the Delta's problems. This list is not exhaustive; a near-infinite number of alternatives exist for managing the Delta. However, these nine alternatives allow us to explore a variety of very different approaches in light of recent understanding of the dilemmas, vulnerabilities, and possibilities for Delta water and land management. Some of these alternatives have been under consideration at various times in the past; others are relatively new. Most seek a "soft landing" from the Delta's current severe disequilibrium and vulnerability.

Three of these alternatives would maintain the Delta as a freshwater body, either by relying on current strategies or by building stronger systems. A second group of alternatives would manage the Delta as a more complex and fluctuating mosaic of uses, supporting water supply exports with peripheral or through-Delta aqueducts. A final group would reduce overall dependence on the Delta, or potentially abandon the Delta altogether. All nine alternatives are outlined below.

Freshwater Delta Alternatives

All three freshwater Delta alternatives would aim to maintain the Delta as a homogeneous freshwater body, continuing policies begun in the 1930s. Levees, outflows, and perhaps barrier structures would be the primary way to control Delta salinity.

1. **Levees as Usual.** The current levee-intensive system would be maintained at recent levels of effort or modestly upgraded to meet federal standards for agricultural levees. Water exports would continue to be pumped through the Delta. Levee failures would occur with increasing frequency.

2. **Fortress Delta.** "Whatever it takes" investments would be made to support or fix levees deemed strategically important for urban areas, infrastructure, and water supply exports. To contain costs, the total length of the levees in the system would be shortened, reconfiguring some islands. Lower-reliability levees (mainly in the interior of the Delta) would be allowed to fail.

3. **Seaward Saltwater Barrier.** A permanent or movable barrier would be erected at the western edge of the Delta. This is one of the oldest and most extreme proposals for keeping salt water at bay, but it has recently reemerged because Dutch engineers have suggested the construction of a large movable barrier, similar to the Maeslant storm surge barrier that protects Rotterdam in The Netherlands.

Fluctuating Delta Alternatives

In all three of these alternatives, environmental conditions, especially salinity, would be allowed to fluctuate in the western Delta to improve habitat conditions for native fish species. Urbanization would be possible along the Delta's periphery behind strong levees.

4. **Peripheral Canal Plus.** An aqueduct would be constructed from the vicinity of Hood, on the Sacramento River, south along the Delta's eastern edge, sending water exports to Clifton Court Forebay. This would allow water exports to circumvent the Delta and yet continue to meet the Central Valley Project and State Water Project intakes that send water to other regions of the state. This proposal augments the traditional peripheral canal proposals with special operations,

investments, and activities for environmental and other in-Delta land and water uses (hence the "plus").

5. **South Delta Restoration Aqueduct.** This aqueduct would be similar to the peripheral canal mentioned above, but its major outlet would enter the lower San Joaquin River. These supplemental freshwater flows would resolve various water quality and flow problems of the lower San Joaquin River and the southern Delta while improving the quality of water exports and reducing entrainment of native fish at the pumps. Some flows could be channeled into a wetland and flood bypass channel through the southern Delta, contributing to improved habitat and agricultural water quality. In-Delta investments would be made for environmental and other in-Delta uses.

6. **Armored-Island Aqueduct.** By armoring select islands and cutting off or tide-gating various channels within the central-eastern Delta, a major, semi-isolated freshwater conveyance corridor for water exports would be created. Various versions of this approach have been considered since the 1950s.

Reduced-Exports Alternatives

These alternatives rely neither on new Delta export facilities nor on levees. However, they imply an ability to greatly modify the pattern and quantity of Delta exports.

7. **Opportunistic Delta.** Only opportunistic seasonal exports would be allowed, during times of high discharge of fresh water from the Delta (generally winter and spring). Export pumping capacities would be expanded to accommodate these high pumping periods, and some surface storage within and near the Delta may be built. Salinity levels would fluctuate in the western Delta, and many islands would eventually become flooded. Urbanization would be possible along the Delta's periphery, behind strong levees.

8. **Eco-Delta.** The Delta would be managed as a single, unified entity to favor key Delta aquatic and terrestrial species. Water extraction, transportation corridors, and other functions would be maintained as long as they do not interfere with rehabilitation goals. Some

water exports would occur but less than in the Opportunistic Delta alternative.

9. **Abandoned Delta.** A planned, multidecade retreat from the Delta would occur, with the phasing out of much of the Delta's farm economy. Water exporting agencies would transition to alternative water sources and would increase water use efficiency.

Our evaluations of these alternatives suggest some promising solutions. A summary of our evaluations appears in Table S.1, along with a summary of our rationale. The intent of our analysis is to eliminate unpromising long-term directions for the Delta and point to some promising approaches, focusing the limited available attention, talent, and resources on those more likely to be successful over time. However, detailed knowledge and analysis will be needed before the identification of a single "best" alternative can be justified.

We find that the first three alternatives, which strive to preserve the Delta as a homogeneous freshwater body, feature unpromising environmental performance at great financial expense, even though some of them would secure substantial quantities of fresh water for export and use within the Delta. In particular, the current approach to managing the Delta—with moderate reinforcement of existing levees and net Delta outflows to keep the Delta fresh—prolongs its risks and vulnerabilities, which are likely to increase over time. Temporary or permanent in-Delta improvements for agricultural and urban land users do not overcome these drawbacks.

The second set of alternatives, which allow for local specialization and variability in the Delta, seem promising and worthy of more detailed development and consideration. These alternatives are built around very different approaches for supporting water exports. In-Delta agricultural and urban users could both see benefits from levee strategies within these alternatives. Although elements of these alternatives will be familiar to many who know something about Delta water policy and politics, each one has some fundamental differences from earlier proposals.

The final set of alternatives modifies current export policies to gain the flexibility to achieve other objectives. At the extreme is the abandonment of the Delta for most purposes. The argument for this strategy is that if the Delta is unreliable and vulnerable, then it might be best to reconfigure state

Table S.1

Summary Evaluation of Alternatives

Alternatives	Summary Evaluation	Rationale
Freshwater Delta		
1. Levees as Usual—current or increased effort	Eliminate	Current and foreseeable investments at best continue a risky situation; other soft landing approaches are more promising; not sustainable in any sense
2. Fortress Delta (Dutch standards)	Eliminate	Great expense; unable to resolve important ecosystem issues
3. Seaward Saltwater Barrier	Eliminate	Great expense; profoundly undesirable ecosystem performance; water quality risks
Fluctuating Delta		
4. Peripheral Canal Plus	Consider	Environmental performance uncertain, but promising; good water export reliability; large capital investment
5. South Delta Restoration Aqueduct	Consider	Environmental performance uncertain but more adaptable than Peripheral Canal Plus; water delivery promising for exports and in-Delta uses; large capital investment
6. Armored-Island Aqueduct	Consider	Environmental performance likely poor unless carefully designed; water delivery promising; large capital investment
Reduced-Exports Delta		
7. Opportunistic Delta	Consider	Expenses and risks shift to importing areas; relatively low capital investment; environmental effectiveness unclear
8. Eco-Delta	Consider	Initial costs likely to be very high; long-term benefits potentially high if Delta becomes park/open space/endangered species refuge
9. Abandoned Delta	Eliminate	Poor overall economic performance; southern Delta water quality problems; like Alternative #1, without benefits

water, environmental, and land use policy to minimize the importance of this unreliable partner. However, we find that the environmental outcome of abandoning the Delta would be poor, because the Delta would not return to anything like its pre-European condition. Moreover, the economic costs for agricultural and other water users would be extreme, on the order of $1.2 billion per year. However, in this group of options, the alternatives that alter export patterns to add fluctuations and improve environmental performance show some promise and merit further consideration.

Adapting to Change

No alternative will be ideal from all perspectives, and some would preclude certain current uses of the Delta entirely. Our analysis suggests that alternatives seeking to maintain the entire Delta as a freshwater system—along the lines of the current levee-centric policy—are incompatible with giving the Delta's native species a fighting chance to survive and prosper. The levee-dependent freshwater alternatives are also the least compatible with the drivers of change currently acting on the Delta, including land subsidence (sinking land elevations), sea level rise, regional climate change, and increased seismic risk, all of which are increasing the risk of levee failure.

Changes in the Delta will have significant costs and cause some dislocations. However, most users of Delta services have considerable ability to adapt economically. As a result, costs and dislocations, if properly managed, should be modest from a statewide perspective. Mitigation should be used to ease adjustment costs. Because they have nowhere else to go, the most vulnerable users of the Delta are those native species that rely on it for survival.

New Ideas for Managing the Delta

Although our analysis draws on the long history of thinking about management options for the Delta, it includes several relatively new ideas.

- **Creating localized Delta specialization.** Traditionally, policymakers have sought to treat the entire Delta homogeneously. Allowing different parts of the Delta to specialize in particular functions or services may make for greater overall sustained performance for all, or almost all, purposes. Spatial and temporal variability in flows, water quality, and habitat was common in the pre-European Delta.

- **Establishing a western Delta fluctuating-salinity ecosystem.** Western Delta salinity appears to have naturally fluctuated more in the past than it does now; reintroducing this fluctuation in parts of the western Delta might benefit native and desirable alien species.

- **Using peripheral areas, such as Suisun Marsh and Cache Slough, to bring back desirable natural conditions that existed in the Delta historically.** These are especially promising examples of locations that could serve valuable environmental functions.

- **Allowing the urbanization of some Delta lands.** Local land use pressures, access to major transportation and employment centers, and financial opportunities make urbanization of some Delta lands seemingly inevitable, despite high risks of flooding. Urbanization has significant potential to contribute financially and politically to solving problems in the Delta. Careful regulation should be able to provide sufficient flood protection and prevent urbanization from unreasonably interfering with environmental functions.

- **Building a Sacramento–San Joaquin Canal (Alternative #5).** Such a canal would supplement lower San Joaquin River flows with Sacramento River water to provide water near export pumps. It would simultaneously improve lower San Joaquin River and southern Delta freshwater quality and availability. This canal would provide larger supplemental flows to the San Joaquin River than earlier peripheral canal proposals.

- **Creating a San Joaquin River marsh and flood bypass.** As part of the Sacramento–San Joaquin Canal alternative, such a system would provide additional habitat for fish and wildlife, water quality improvements for southern Delta farmers, and flood bypass capacity for the lower San Joaquin River.

- **Managing expectations and providing mitigation alternatives.** It is unlikely that any Delta solution can satisfy all Delta interests in terms of water and land use. This approach differs from the underlying assumption of CALFED that all Delta interests could "get better together." Stakeholders whose land and water interests cannot be directly satisfied may be compensated by financial or other means. Even with such mitigations and compensations, one cannot reasonably expect universal satisfaction.

Conclusions

This report has five major conclusions:

1. The current management of the Delta is unsustainable for almost all stakeholders. The combined effects of continued land subsidence, sea level rise, increasing seismic risk, and worsening winter floods make continued reliance on weak Delta levees imprudent and unworkable over the long term.
2. Recent improvement in the understanding of the Delta environment allows for more sustainable and innovative management. Seeing the Delta as a functioning ecosystem with fluctuating flows and salinity, as it once was, allows us to think of new solutions to the Delta's problems.
3. Most users of Delta services have considerable ability to adapt economically to risk and change. Water and land users have a wide variety of adaptive responses, which, although sometimes costly, do allow them to adjust. Moreover, users of the Delta also have a history of responding to change; many are already adapting in anticipation of worsening problems in the Delta.
4. Several promising alternatives exist to current Delta management. The situation is far from hopeless. A sustainable, prosperous Delta economy and society can be built while providing water and other services statewide.
5. Significant political decisions will be needed to make major changes in the Delta. Incremental, consensus-based solutions are unlikely to prevent a major ecological and economic catastrophe of statewide significance.

Recommendations

We recommend several actions and activities.

1. **Create a technical track for developing Delta solutions.** Most recent attempts to solve the Delta's problems have been politically driven. Agencies and other stakeholders have sought to negotiate solutions based on what is politically acceptable, but this approach has not led to acceptable or workable solutions. Despite improvements in our understanding of the Delta ecosystem and the economy of California, little in the way of new solutions or approaches to the Delta has been developed or proposed. Now we are all "getting worse together." The political track of any Delta solution is necessary, but it can be better informed by a technical track, which can develop new solutions and adapt older solutions to current and future conditions. There is strong historical precedent for this: In 1911, the California Debris Commission provided such a service, suggesting effective long-term solutions for the Sacramento Valley flood control problems.

2. **Establish an institutional framework to support the development of solutions and to bring scientifically and economically promising alternatives to the attention of political authorities.** This activity needs to take a long-term view and avoid crisis-driven responses to short-term political thinking. It should have some political independence, an appropriately sized budget, the technical capability to creatively and competently explore and eliminate alternatives, and the management capability to direct multidisciplinary research and development. CALFED was supposed to have these abilities, but its direction, funds, and energy became dissipated in politics and the effort to please all stakeholders. Current technical efforts examining both the pelagic organism decline and the risks to Delta levees focus rather narrowly on specific aspects of the Delta's problems. Current policy efforts—including the Delta Vision process—lack a substantial technical component. Technical and policy endeavors need some independence within a larger framework.

3. **Launch a problem-solving research and development program.** The science effort regarding the Delta needs an overhaul. The Delta is a multidisciplinary problem, not a research topic with a single focus. Much past research on the Delta and its problems has been associated with agency data collection or basic academic and disciplinary research. A directed problem-solving research and development

program aimed primarily at developing and informing the analysis of promising solutions is needed. This program would include some basic research, but most effort would be aimed at developing and evaluating solutions. Ecosystem adaptive management experiments (supported by quantification and computer modeling), levee replacement, island land management, flood control, and integrative system design would receive greater attention in a problem-solving framework.

4. **Consider the Delta's water delivery problems in a broad context.** The foremost physical problem in the Delta is delivery of fresh water through or around the Delta. And some promising solutions exist. Potential options extend beyond the peripheral canal. However, physical solutions for water delivery must be accomplished in the broader context of developing a more sustainable Delta environment.

5. **Eliminate some solutions to the Delta's water delivery problems from further consideration.** To reduce investments of scarce time, expertise, and resources in evaluating Delta alternatives, some unpromising options should no longer be considered. These include the current levee-centric approach, the building of a downstream physical barrier to seawater, the large expansion of on-stream surface water storage, and the idea of ending all water exports. These are unreasonable solutions and they perform so poorly in economic and environmental terms as to be nonviable.

6. **Approach the Delta as a diverse and variable system rather than as a monolith.** A diversified and variable Delta by design is likely to perform better than the freshwater Delta that has been artificially maintained over the last 60 years. Better solutions are likely to emerge if the Delta is not treated homogeneously. Historically, the Delta naturally contained diverse habitats that varied across years, seasons, and tidal cycles in terms of salinity, water velocity, water clarity, elevation, and other physical habitat conditions. Reintroducing and extending this diversity, by specializing parts of the Delta for wildlife habitat, agriculture, urbanization, recreation, water supply, and other human purposes seem promising.

7. **Give direct beneficiaries primary responsibility for paying for Delta solutions.** Public funds, such as those raised through general

obligation bonds, should be reserved for the truly public components of a Delta investment program, such as ecosystem restoration and mitigation for those who lose out. Failure to develop an effective funding mechanism will result in financial catastrophes for state and local interests in the future, especially in the wake of a natural disaster.

8. **Establish mitigation and compensation mechanisms to support the implementation of any alternative.** Not all parties will get what they want or what they have been used to getting from the Delta. In some cases, providing money or alternative land might compensate for changing or eliminating uses of water or land that would hinder broad progress.

9. **Create stronger regional and statewide representation in Delta land use decisions.** The current institutional fragmentation of land use authorities in the Delta fosters piecemeal decisionmaking that will compound flood risks, irreversibly destroy valuable wildlife habitat, and deteriorate water quality. Regional and statewide interests should be more forcefully represented in Delta land use decisions, to protect the value of the Delta both for the region's residents and for the broader public. The Delta needs a strong regional permitting authority, along the lines of the San Francisco Bay Conservation and Development Commission or the Coastal Commission.

10. **Make essential emergency preparedness investments.** Although it is premature to choose a long-term solution for the Delta without further technical investigation, California can take steps in the short term. All agencies relying on Delta waters should develop extended export outage plans through regional interties, water sharing arrangements, and other measures. Other infrastructure providers also need contingency plans. A program for the rapid repair of critical levees, such as the one launched in 2006, and emergency flood response plans are key.

11. **Implement a "no regrets" strategy for the Delta.** First, given the urbanization pressures on the Delta, policy decisions are needed to establish an improved regional governance structure, institute a program to set aside or purchase key habitat, and create adequate, coherent flood control guidelines for urbanizing lands. Second, to avoid costly expenditures for islands that are of low strategic value, it

makes sense to develop a "do not resuscitate" list in the event of levee failure. Third, to improve habitat conditions for the delta smelt and other pelagic species in the short term, restoration actions should be initiated in the Suisun Marsh and Cache Slough regions.

Forging a New Path Forward

The Delta's many problems have sparked a crisis in confidence among its many stakeholders. The CALFED process, which has been responsible for crafting solutions in the Delta since the mid-1990s, is now widely perceived as having failed to meet its objectives. That process was forged under the threat of new federal water quality standards for the Delta. CALFED's failure lay in the course chosen for crafting solutions: favoring political consensus over making tough choices among alternatives and assuming that taxpayer largesse would foot any bill. The question going forward is whether the crisis in the Delta can spur stakeholders and the state to action with a new strategy that accepts the inevitability of both winners and losers. The future of this unique ecosystem and regional resource and of the state's water supply system all depend on the answer. All Californians are likely to see benefits (and costs) from a comprehensive long-term solution. Otherwise, we will all see only costs.

Contents

Figures

Tables

Preface and Acknowledgments

"No question is ever settled until it is settled right."

Ella Wheeler Wilcox (1850–1919)

This report is the result of a true multidisciplinary examination of the Delta. The authors include two economists (Ellen Hanak and Richard Howitt), two engineers (William Fleenor and Jay Lund), a biologist (Peter Moyle), and a geologist (Jeff Mount). Many readers will recognize most of the authors from their past involvement in a wide range of water management and policy issues in California. The report is based on discussions and investigations conducted over a nine-month period, including extensive conversations with additional agency and nongovernmental organization experts in the Delta, specializing in policy, hydrodynamics, and ecosystems. The result is the consensus presented here.

Several of the chapters and findings would be much weaker without the diligent efforts of two Ph.D. students, Stacy Tanaka and Marcelo Olivares, who conducted the CALVIN and DAP modeling studies, respectively. They are the lead authors of the two appendices on these modeling activities.

We wish to thank those who contributed by participating in two technical workshops. The first workshop focused on hydrodynamic issues in the Delta and the second focused on ecosystem performance under different Delta scenarios. Participants in one or both of these workshops included Bill Bennett, Jon Burau, John Cain, John DeGeorge, Chris Enright, Marianne Guerin, Bruce Herbold, Paul Hutton, Richard Rachiele, K. T. Shum, Pete Smith, Tara Smith, and Ted Sommer. We also thank the many people who generously agreed to be consulted over the course of the research through in-person meetings or telephone conversations (Appendix B).

Very helpful reviews of the first draft of this report were provided by Elisa Barbour, Alf Brandt, Jon Burau, Kamyar Guivetchi, Jon Haveman, Bruce Herbold, B. J. Miller, Laura King Moon, Mary Nichols, Michael Teitz, and David Zilberman. Their input led to considerable improvements.

We alone are responsible for any remaining errors and for all interpretations of the material presented herein.

We also wish to thank Gary Bjork and Lynette Ubois (PPIC) and Patricia Bedrosian (RAND) for editorial support and Janice Fong (UC Davis Geology Department) for designing the maps of the Delta used in this report.

The Public Policy Institute of California financially supported the involvement of Stacy Tanaka and Marcelo Olivares as well as the other UC Davis activities leading to this report through a grant to the John Muir Institute of the Environment at UC Davis. We are extremely grateful for this financial support and for the clerical and office support of the John Muir Institute for the Environment and its Watershed Science Center on the campus.

Acronyms and Abbreviations

ACWA	Association of California Water Agencies
BDCP	Bay-Delta Conservation Plan
CAL EPA	California Environmental Protection Agency
CALAG	California Agricultural Model
CALFED	state and federal program for the San Francisco–San Joaquin Bay Delta
CALSIM	water resources simulation model
CALVIN	California Value Integrated Network Model
CCC	Contra Costa Canal
CCWD	Contra Costa Water District
CDEC	California Data Exchange Center
CES	constant elasticity of substitution
CESA	California Endangered Species Act
cfs	cubic feet per second
CVP	Central Valley Project
CVPIA	Central Valley Project Improvement Act
CVPM	Central Valley Production Model
DAP	Delta Agricultural Production Model
DRMS	Delta Risk Management Study
DWR	Department of Water Resources
EBMUD	East Bay Municipal Utilities District
EC	electrical conductivity
ESA	Endangered Species Act (federal)
EWA	Environmental Water Account
GCM	General Circulation Models
HCP	Habitat Conservation Plan
maf	million acre-feet
MNDO	monthly net Delta outflow
MWDSC	Metropolitan Water District of Southern California
NCCP	Natural Communities Conservation Plan
OCAP	operating criteria and plans
PC	peripheral canal
PG&E	Pacific Gas & Electric

PL	Public Law
POD	pelagic organism decline
ppt	parts per thousand
ROD	Record of Decision
SDRA	South Delta Restoration Aqueduct
SFBCDC	San Francisco Bay Conservation and Development Commission
SWAP	Statewide Water and Agricultural Production Model
SWP	State Water Project
SWRCB	State Water Resources Control Board
taf	thousand acre-feet
USBR	U.S. Bureau of Reclamation
USEPA	U.S. Environmental Protection Agency
USFWS	U.S. Fish and Wildlife Service
USGS	U.S. Geological Survey
VAMP	Vernalis Adaptive Management Program

Glossary: Words and Phrases

Anadromous fish species—Fish that live in ocean water and move inland to spawn, such as salmon.

Beneficiary pays principle—The principle that financial responsibility for a project is apportioned among those who benefit from it.

Consumptive water use—Diversions of water withdrawn but not returned downstream.

Dendritic channels—Branching or branchlike water channels, as found in the Delta before land reclamation.

Fish entrainment—The drawing of fish or fish larvae into pumps or water diversions.

Fishery—The organized capture of fish for sport or commercial purposes.

Groundwater banking—The managed storage of water in underground aquifers.

Inflows—Natural or managed flows of water into a particular location.

Interties/intertied water system—Connections between water conveyance facilities. An intertied water system such as California's is quite flexible and cost effective in accommodating water shortages or malfunctions at specific water management facilities.

Land subsidence—The sinking of lands caused by compaction, oxidation of peat soils, and wind erosion. Many Delta islands have subsided (mostly from oxidation and erosion) to the point where they now lie many feet below sea level.

Mitigation—An action intended to moderate some effects of other activities. For instance, flood management agencies often make one-time payments (known as "flood easements") to property owners in areas that will be allowed to flood periodically to help cover the costs of flooding.

Outage plans—Plans for the unavailability of a resource or facility (such as a loss of water or electricity resulting from the failure of a pump or transmission line).

Outflow— Flows of water going away from a particular location.

Pelagic fish species—Fish that live their whole life in open water, above the bottom. Within the Delta, this category includes the delta smelt, long-fin smelt, and striped bass.

Planform—Two-dimensional patterns on the land, as seen from above or on a planning map.

Reclamation—The diking and draining of swamp lands. Most "reclaimed" Delta lands are used for agriculture, although some lands are used for wildlife habitat and urban development.

Residence time—The length of time water remains within a particular channel or area.

Salinity—The concentration of salt in water. As a rough guide, sea water is 35 ppt (grams per liter) and fresh water is less than 3.0 ppt. Drinking water is less than 1.0 ppt.

Tidal excursions—The mixing of waters caused by daily tidal movements in and out of an estuary.

Water exports—Generally refers to water used somewhere other than its area of origin. Direct Delta exports refers to water from Delta watersheds that is sent to points south and west of the Delta. Indirect exports, also known as upstream diversions, refers to water diverted from the Delta watersheds (mainly in the Sacramento Valley and on the east side of the San Joaquin Valley) before it reaches the Delta.

Water diversions—The withdrawal of water from a water body, some of which might be returned downstream after use.

Water scarcity—When water deliveries are less than desired. Scarcity is often managed by price, rationing urban water use, fallowing some farmland, or curtailing recreational activities.

Water transfers—The exchange, leasing, or permanent sale of the rights to use a particular amount of water from a particular source. Such transfers occur through a "water market," usually involving local and regional water agencies and often state and federal agencies.

Water year—California's water year begins on October 1st, the beginning of the rainy season, and ends on September 30th.

1. Introduction

"People seldom see the halting and painful steps by which the most insignificant success is achieved."

Anne Sullivan (1866–1936), American Educator of the Deaf, Blind

The Sacramento–San Joaquin Delta is the hub of California's water system, home to a unique ecosystem and to a productive agricultural and recreational economy. Strategies to manage the Delta that would satisfy competing interests have been discussed and debated for almost 100 years, at times leading to acrimonious divisions between Northern and Southern California, environmental and economic interests, and agricultural and urban sectors. Recently, the Delta has again taken center stage in debates on California water policy. Research and actual levee failures have exposed the New Orleans–level fragility of 1,100 miles of levees, on which both Delta land uses and water supply systems currently depend. In addition, dramatic declines have occurred in the population of several fish species that depend on the Delta. Furthermore, the institutional framework known as CALFED—a stakeholder-driven process established in the mid-1990s to mediate conflict and to "fix" the problems of the Delta—is facing a crisis of confidence. As the CALFED truce erodes, lawsuits are beginning to fill the gaps left by a lack of consensus on management strategies and options.

For the past 70 years, the state's policy has been to maintain the Delta as a freshwater system through a program of water flow regulation, supported by the maintenance of agricultural levees. This approach now appears near or past the end of its useful life, given the deteriorating state of the Delta's ecosystem and levees as well as the rising consequences of levee failure. This report is about a search for solutions to Delta problems. We do not pretend to offer the definitive solution; 100 years of history would argue that such a solution is unlikely. Indeed, it may be that different Delta strategies are appropriate for different periods of California's development. Instead, our aim is to launch a serious, scientific search and comparison of potential long-term solutions for the coming decades.

What Is the Delta?

The Delta is a web of channels and reclaimed islands at the confluence of the Sacramento and San Joaquin Rivers. It forms the eastern portion of the wider San Francisco Estuary, which includes the San Francisco, San Pablo, and Suisun Bays, and it collects water from California's largest watershed, which encompasses roughly 45 percent of the state's surface area and stretches from the eastern slopes of the Coastal Ranges to the western slopes of the Sierra Nevada. It resembles other deltas of the world in that it is at the mouth of rivers, receives sediment deposits from these rivers, and was once a vast tidal marsh. The Sacramento–San Joaquin Delta is fundamentally different from other delta systems, however, in that it is not formed primarily by the deposition of sediment from upstream. Instead, it is a low-lying region where sediment from the watershed commingled with vast quantities of organic matter deposited by tules and other marsh plants. For some 6,000 years, sediment accumulation in the Delta kept up with a slow rise in sea level, forming thick deposits of peat capped by tidal marshes. A century and a half of farming has reversed this process, creating artificial islands that are mostly below sea level, protected only by fragile levees. Today, those who drive through the Delta see mainly huge tracts of flat, prosperous farmland intersected by narrow channels populated by recreational boaters.

Geographically, the area known as the "Legal Delta" lies roughly between the cities of Sacramento, Stockton, Tracy, and Antioch (Figure 1.1). It extends approximately 24 miles east to west and 48 miles north to south and includes parts of five counties (Sacramento, San Joaquin, Contra Costa, Solano, and Yolo). At its western edge lies Suisun Marsh, an integral part of the Delta ecosystem. At its southern end, near Tracy, motorists pass over two major pieces of California's water infrastructure—the Delta-Mendota Canal and the California Aqueduct. These and several smaller aqueducts, built between the 1930s and the 1960s, deliver water from Northern California rivers to cities and farmland in coastal and Southern California and the San Joaquin Valley. The Delta is considered the hub of the state's water supply because it is used as a transit point for this water. This role has significantly influenced Delta management policies, which aim to keep Delta water fresh.

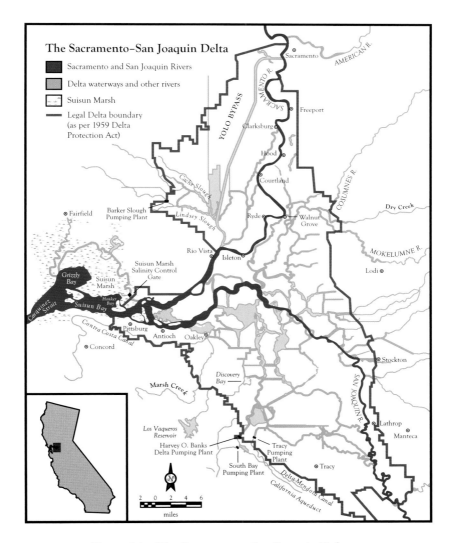

Figure 1.1—The Sacramento–San Joaquin Delta

Today, the Delta supports a highly modified ecosystem. It resembles the Delta of the past only in that some of the original species, such as delta smelt and Chinook salmon, are still present, albeit in diminished numbers. Invasive organisms, from plants to fish, now dominate the Delta's steep-sided channels and long-flooded islands (mainly Franks Tract and Mildred

Island). Most of the native fish either migrate through the Delta (e.g., Chinook salmon, steelhead, splittail) or move into it for spawning (delta smelt and longfin smelt). Resident native fish are present mainly in areas strongly influenced by flows of the Sacramento River. Although the past decade has witnessed some improvements in salmon populations (often grouped under the heading "anadromous" because they live in ocean water and move inland to spawn), the delta smelt and other open-water or "pelagic" species have sharply declined in recent years. Habitats in marshlands and along the banks of rivers ("riparian" areas) have been reduced to small remnants in the Delta, although agricultural lands are important winter foraging areas for sandhill cranes and various waterfowl (Herbold and Moyle, 1989).

Why the Delta Matters to Californians

Most Californians rely on the Delta for something, whether they know it or not. Approximately 50 percent of California's average annual streamflow flows to the Delta. Most Californians drink water that passes through the Delta, and most of California's farmland depends on water tributary to the Delta.[1] And, increasingly, people are building their homes in the Delta, perhaps not realizing the risks to their property and lives from living near or below sea level behind undersized and poorly maintained levees. Table 1.1 summarizes the many ways in which California's regions receive services from the Delta.

Clearly, the Delta is not merely a hub for water supply. It is also a center for important components of California's civil infrastructure. The electricity and gas transmission lines that crisscross the region serve many parts of the state. The Delta is also used for the underground storage of natural gas to accommodate peak wintertime demands. Furthermore, the Delta hosts several transportation lines. California's major north-south highway (I-5) goes through its eastern edge, and two commuter routes— SR 4 and SR 12—cross its southern and central portions, respectively (Figure 1.2). Several rail lines pass through the heart of the Delta, as do the deepwater shipping channels leading to the ports of Stockton and Sacramento. In addition, aqueducts and canals conveying water to several

---

[1]See Chapter 6 for details on water use by region.

Table 1.1

Services Supplied by the Delta Region to Areas of California

Delta Service	Benefiting Region			
	North of Delta	In-Delta	South of Delta	West of Delta
Agricultural land use		√		
Urban land use		√		
Ecosystem nutrients and support		√		√
Migration routes for salmon and other fish	√	√	√	√
Water supply	√	√	√	√
Recreation (boating, fishing, hunting, ecotourism)	√	√	√	√
Commercial shipping	√	√	√	√
Natural gas mining and power generation	√	√	√	√
Electricity and gas transmission and gas storage	√	√	√	√
Road and rail connections	√	√	√	√
Salt, waste, and drainage disposal	√	√	√	
Water supply right-of-way				√

NOTES: North of Delta includes the Sacramento Valley. In-Delta includes Delta Island users. South of Delta includes Southern California and the eight-county San Joaquin Valley. West of Delta includes the San Francisco Bay Area (including Contra Costa County).

west-of-Delta water utilities—including the East Bay Municipal Utilities District and the Contra Costa Water District—also pass through parts of the Delta. And two power plants are at the Delta's western edge, in Antioch and Pittsburg.

In addition to civil infrastructure, the Delta also provides crucial habitat, and many of California's fish species live in or migrate through it. Moreover, the Delta is valued for its aesthetic appeal and for its support of recreational activities. Its proximity to population centers in the Bay Area, Sacramento, and the northern San Joaquin Valley makes it an attractive destination for boating, fishing, hunting, and ecotourism. The Delta's 635 miles of boating waterways are served by 95 marinas containing 11,700 in-water boat slips and dry storage for 5,500 boats. In 2000, there were

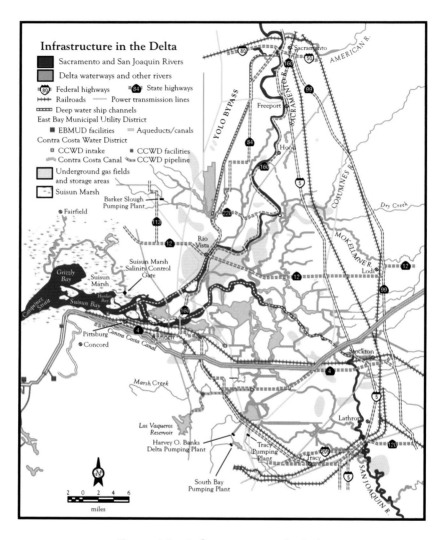

Figure 1.2—Infrastructure in the Delta

an estimated 6.4 million boating-related visitor-days, with 2.13 million boating trips. Recreational boating is expected to grow to 8.0 million visitor-days by 2020 (Department of Boating and Waterways, 2002). Fishing is also a popular activity (Plater and Wade, 2002), as is duck hunting in the Suisun Marsh.

The Delta also serves as a vast drainage area for polluted agricultural and urban runoff. This runoff contains a variety of surplus and residual pesticides and nutrients, in addition to contaminants leached from the soils of specific regions. Drainage from within the Delta contains dissolved organic compounds from the islands' peaty soils, which increase water treatment costs and drinking water quality risks. Sacramento Valley drainage includes mercury and other wastes from historic mining activities, and San Joaquin Valley agricultural drainage includes salts originating in the soils in the Valley's west side and in irrigation water. Retaining such wastes locally would cause great expense and impairment within the source regions, but allowing them to flow into the Delta creates water quality problems for human and environmental uses within the Delta and beyond.

Finally, the Delta provides land. Until recently this land had been used predominantly for agriculture. Today, however, the Delta's land, as well as its water, has come into greater demand for urban, environmental, and recreational uses.

The Delta in Crisis

Concerns for the continued provision of services from the Delta involve several issues:

- Land subsidence, sea level rise, and changes in climate make Delta levees increasingly vulnerable to failure from earthquakes, floods, and other causes.
- Endangered species and fisheries have continued to decline in the Delta and disruptive nonnative species continue to invade.
- Delta water quality remains at risk from salts entering from the ocean and the San Joaquin Valley's agricultural drainage as well as from pesticides and metals coming from agricultural and urban lands.
- Regional population and economic growth has increased pressure to urbanize Delta lands near major transportation routes and urban centers. This "hardening" of Delta lands simultaneously raises the costs of flood risks and reduces the flexibility of land management options.

Awareness of these issues has intensified over the past two years, leading many to question the viability of current policies for the Delta. Indeed, by several key criteria, the Delta is now widely perceived to be in crisis. One dimension of the crisis is the health of the levees. The devastating effects of Hurricane Katrina on levees in New Orleans galvanized public attention on the fragility of the Delta's levee system, where close calls occur with some frequency; for example, a Jones Tract levee broke in June 2004. Recently, the Department of Water Resources (DWR) has publicized the economic consequences of a catastrophic levee failure caused by a large earthquake. One scenario, which envisaged 30 levee breaches and 16 flooded islands, predicted that water exports would be cut off for several months, that shipping to the Port of Stockton would be cut off, and that there would be disruptions of power and road transportation lines (Snow, 2006). The total cost to the economy, over five years, was estimated at $30 billion to $40 billion. A similar study of a 50-breach scenario, focusing only on the costs to water users, put the annual costs of a shutdown at the pumps at $10 billion (Illingworth, Mann, and Hatchet, 2005).

A second aspect of the crisis is the health of Delta fish species. In the fall of 2004, routine fish surveys registered sharp declines in several pelagic species, including the delta smelt, a species listed as threatened under the Endangered Species Act. Subsequent surveys have confirmed the trend, raising concerns that the smelt—sometimes seen as an indicator of ecosystem health in the Delta—risks extinction if a solution is not found quickly (Figure 1.3).

The third dimension of the crisis is institutional. The CALFED process that has been responsible for coordinating Delta solutions since the mid-1990s has faced serious problems since late 2004. CALFED's failure to anticipate funding and disagreements among stakeholders on some key elements of its program has contributed to a loss of confidence in this institutional framework (Little Hoover Commission, 2005). Since the summer of 2006, the California Bay Delta Authority—the body responsible for coordinating CALFED activities—has been operating out of the Resources Agency, without an independent budget. Thus, the strong leadership and financial resources needed to address the Delta's problems are currently lacking.

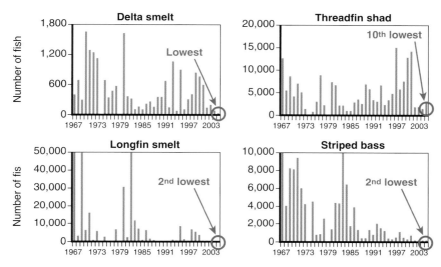

SOURCE: California Department of Fish and Game.
NOTES: Graphs report the indices for the fall midwater trawl. Circles indicate the rank of indices in 2005. For delta smelt, longfin smelt, and striped bass, the recent indices represent low points in long-term declines of their populations.

Figure 1.3—Fall Abundance Indices for Several Pelagic Fish Species in the Delta, 1967–2005

Responding to the Crisis

Recognition of the crisis in the Delta has led to appeals to pursue a number of very different management strategies. The collapse of Delta fish populations has prompted some environmentalists to call for cutbacks in water exports. Meanwhile, two main proposals have surfaced for dealing with levee instability: massive investments in the levee system to reduce the risk of failure (creating, in a sense, the "Fortress Delta" we describe below) or construction of a peripheral canal at the Delta's eastern edge, to protect water exports from what many now view as unacceptable risks associated with direct Delta exports. The resurgence of a peripheral canal proposal is significant, because it is a solution that has deeply divided Californians in the past. Strong majorities of Northern California and San Joaquin Valley voters—concerned over the canal's environmental effects, its potential to export too much water south, and the proposed allocation of costs—succeeded in defeating a peripheral canal proposal in a statewide referendum in 1982. When the CALFED process was launched in the mid-1990s to find new solutions to the Delta's ecosystem and water supply issues, feelings were still so raw that the peripheral canal was not considered an acceptable option.

These proposals have largely emerged from stakeholder groups, and none provide fully fleshed-out plans to address the Delta's woes. To date, the only concrete response from Sacramento, supported by both the governor and the legislature, has been to put more state funds into shoring up Delta levees, which were relatively neglected under CALFED.[2] State budget allocations for levee repairs were increased significantly in 2006, and two bond measures passed in November 2006 ballot allocate additional funds for flood control in the Delta. However, there is as yet no broad plan for responding to the crisis in the Delta, including how the bond funds should be spent.

Such a plan may emerge from several efforts now under way or envisioned. Recently, two focused scientific studies have been launched by government agencies. Since the summer of 2005, a multiagency task force has been examining the causes of the pelagic organism decline (the "POD" study). In the spring of 2006, the Department of Water Resources initiated a two-year "Delta Risk Management Study" (DRMS) to analyze risks to the levee system. Two policy-driven efforts are also beginning. In September 2006, the governor launched a Delta Vision exercise to look at long-term alternatives for the Delta, in conjunction with stakeholders and an independent Blue Ribbon Task Force.[3] Also, as its first phase comes to a close in 2007, the CALFED program must reconsider alternative management strategies to meet its water and environmental goals for the Delta.

The purpose of this report is to provide input into these current processes and into other Delta discussions, by outlining some major issues facing the Delta and initiating a search for long-term solutions. In assessing potential solutions, we purposely take a broader view of the options than

---

[2]In the first four years of the CALFED program, a total of $78 million was spent on levees, only 29 percent of the amount envisaged in the CALFED Record of Decision. Total CALFED spending from all sources was $2.5 billion, 66 percent of the level envisaged (Department of Finance, 2005; CALFED, 2000c).

[3]See Senate Bill 1574 and Executive Order S-17-06, signed on September 28, 2006.

those commonly under discussion in stakeholder circles—namely, the Fortress Delta, the peripheral canal, and the maintenance of the current levee-centric strategy with lower water export volumes.

The task at hand is urgent, and the stakes in the Delta are high. If California fails to develop a viable solution and act on it soon, we risk the loss of native species and significant disruptions of economic activity. Yet there is also a risk that the political process will prematurely close off the consideration of options that could help California make the most of the Delta, while protecting its unique ecosystem and species. Therefore, we seek to contribute to the discussion of the Delta in two ways—first, by describing and evaluating a wide range of strategies for Delta solutions and, second, by pointing out solutions that are not viable and do not merit continued consideration. Time is of the essence, and determining a practical and focused array of options will best serve the interests of all involved in determining the Delta's future.

Crafting Long-Term Solutions for the Delta

Long-term solutions for the Delta will need to consider a wider range of issues than simply which levees to upgrade. To be viable, Delta solutions will need to address four central issues: the salinity of Delta waters, in-Delta land use and water supply, water supply exports, and the Delta ecosystem.

Delta Salinity

With rivers feeding into it and marine bays at its western edge, the Delta is the meeting point for seawater and fresh water within the wider estuary system (Knowles, 2002). Delta salinity has been a major concern since the City of Antioch's 1920 lawsuit against irrigators in the Sacramento Valley, whose upstream water withdrawals reduced freshwater flows into the Delta and increased the salinity at water intakes in the western Delta (Jackson and Paterson, 1977). Salinity affects the potability and taste of urban water supplies, the productivity of farmland, and the viability of different organisms within aquatic ecosystems. For many decades, this issue was discussed in terms of where the salinity gradient— that is, the transition from fresh water to seawater—should be located in the estuary. Since the 1920s, it has been regarded as desirable to maintain the Delta, as much as possible, as a freshwater system, Suisun Bay and

Marsh as brackish water systems, and San Francisco Bay as a marine (saltwater) system. The current regulatory framework for water quality in the Delta rests on this idea. More recent thinking, discussed in Chapter 4, holds that seasonal and interannual variability in much of the estuary may better mimic the natural salinity regime and help limit the extent of invasive species, which tend to prefer waters with little salinity fluctuation. Increasingly, it has been recognized that salinity and other, broader water quality problems in the Delta are compounded by the quality of upstream and in-Delta drainage, with consequences both for urban and agricultural users as well as for fish and wildlife.

Delta Land Use

Land is a central issue for the Delta. Of the Delta's 738,000 acres, roughly two-thirds support agriculture and one-tenth urbanized populations. Although the human population within the heart of the Delta is minimal—limited principally to homesteads and a handful of small "legacy" towns—larger cities such as Stockton and Antioch have long existed on its fringes. The Delta is often thought of as a site of high-value fruit and vegetable farms, but roughly 75 percent of the farmland is actually devoted to lower-value pasture and field crops; in comparison, only 55 percent of farmland statewide is devoted to these uses (Department of Water Resources, 1998). And in recent years, urbanization and recreational use of Delta lands has been on the rise.

Various environmental uses of Delta land already exist, including wetlands, riparian habitat, waterfowl uses, and aquatic habitats. Open water—which results when islands are flooded and submerged—also has environmental use, as well as considerable value for recreation, boating, and shipping. Freshwater storage is another recent suggestion for Delta lands. This freshwater storage plan proposes investing in strengthening internal levees on some Delta islands that have subsided below sea level, allowing them to be filled with water, on a tidal or seasonal time scale, to aid water projects in pumping fresh water from the Delta.[4]

Each of these land uses has different implications for water use, the quality of water required in adjacent channels, drainage quality and

---

[4]One proposal, known as the Delta Wetlands project, is one of five surface storage proposals endorsed by the CALFED program for further consideration (see Chapter 4).

quantity, and economic sustainability. Fortunately, the Delta is large and diverse enough to support a mix of land uses.

Water Exports

Water exports from the Delta are a major cause of controversy. For water users in Southern California, the Bay Area, and the San Joaquin Valley, the reliability and quality of these water supplies are of paramount concern. Yet there are also concerns that export patterns and volumes harm species' health and water quality within the Delta. Many approaches exist for either providing or avoiding this function for the Delta, and numerous options have been proposed over the past century. Even without providing water exports, however, the Delta would still have many serious problems with flooding, land subsidence, degraded habitat, invasive species, and water quality.

Delta Ecosystem

Different parts of the Delta provide habitat for different wild species and their diverse life stages. The mix of salt, brackish, and freshwater marshes as well as upland, riverine, and deepwater habitats affects the abundance and makeup of native and alien species. Therefore, anything that changes the physical Delta changes the biological Delta. Since the 1970s, considerable attention has been paid to the effect of water supply functions on ecosystem functions in the Delta. Initially, this discussion focused primarily on the role of water export pumps at the Delta's southern edge, and on efforts aimed to avoid fish entrainment (the drawing of fish into the pumps). It is now recognized that the same issues of entrainment of fish and invertebrates apply to power plant cooling water and agricultural and urban diversions elsewhere in the Delta. Concerns have also been raised that the total volume and timing of diversions are causing problems for key Delta species by changing the way water flows through the Delta. Given the range of federal and state environmental laws protecting these species, these concerns are legal and political as much as ecological.

Searching for a Soft Landing

In this report, we look for long-term solutions to these chronic, dire, and potentially catastrophic problems. We review a range of alternatives for the Delta—some old and some new—that address these four issues. Rather than focus on crisis management, we consider long-term management strategies, under which Californians can develop and implement a plan to adjust to the Delta of the future. This approach, which we refer to as planning for a "soft landing," differs greatly from how California may need to manage short-term crises in the Delta, or what might be considered a "hard landing." If the state is unfortunate enough to experience a multilevee failure before implementing a long-term plan, effective emergency response will be needed to minimize the costs in terms of water supply and damages to other economic infrastructure. Studies such as DRMS will provide invaluable input into such response plans.

Report Overview

This report develops and explores five major themes:

1. The current Delta is unsustainable for almost all stakeholders.
2. An improved understanding of the Delta environment now allows for more sustainable and innovative management.
3. Most users of Delta services have considerable ability to adapt economically to risk and change.
4. Several promising alternatives exist to current Delta management.
5. Significant political decisions will be needed to make major changes.

The first part of this report focuses on the first three of these themes. Chapter 2 provides a short history of the Delta and draws lessons from past policy interventions. Chapter 3 presents an overview of current problems and future prospects for the Delta in light of the key natural and human drivers of change. Paradigms for understanding and managing the Delta ecosystem are developed in Chapter 4, particularly relating the ecosystem to fluctuating salinity regimes. Chapter 5 focuses on institutional aspects of the current crisis, with a review of stakeholder perspectives. Chapter 6 analyzes the role of Delta water supplies in various regions of California and the ability of water users and the larger water supply system to adjust to changes in Delta water management policies.

The second part of the report turns to an analysis of long-term solutions for the Delta. Chapter 7 presents a range of options and alternatives for managing the Delta. A preliminary assessment of nine alternatives is provided in Chapter 8. Chapter 9 considers various policy issues that will be central to crafting a new Delta framework: principles for financing Delta investments, strategies to provide mitigation for those who may bear a disproportionate share of the costs of particular Delta solutions, and governance issues. Conclusions and recommendations are presented in Chapter 10.

2. The Legacies of Delta History

"You could not step twice into the same river; for other waters are ever flowing on to you."

Heraclitus (540 BC–480 BC)

The modern history of the Delta reveals profound geologic and social changes that began with European settlement in the mid-19th century. After 1800, the Delta evolved from a fishing, hunting, and foraging site for Native Americans (primarily Miwok and Wintun tribes), to a transportation network for explorers and settlers, to a major agrarian resource for California, and finally to the hub of the water supply system for San Joaquin Valley agriculture and Southern California cities. Central to these transformations was the conversion of vast areas of tidal wetlands into islands of farmland surrounded by levees. Much like the history of the Florida Everglades (Grunwald, 2006), each transformation was made without the benefit of knowing future needs and uses; collectively these changes have brought the Delta to its current state.

Pre-European Delta: Fluctuating Salinity and Lands

As originally found by European explorers, nearly 60 percent of the Delta was submerged by daily tides, and spring tides could submerge it entirely.[1] Large areas were also subject to seasonal river flooding. Although most of the Delta was a tidal wetland, the water within the interior remained primarily fresh. However, early explorers reported evidence of saltwater intrusion during the summer months in some years (Jackson and Paterson, 1977). Dominant vegetation included tules—marsh plants that live in fresh and brackish water. On higher ground, including the numerous natural levees formed by silt deposits, plant life consisted of coarse grasses; willows; blackberry and wild rose thickets; and galleries of oak, sycamore, alder, walnut, and cottonwood. Few traces of this earlier plant life remain; agricultural practices and urbanization have cleared most

---

[1]Unless otherwise noted, the discussion in this section draws from Thompson (1957).

17

forested areas and levee upgrading has removed most trees and vegetation from the natural levees.

Before European settlement, the Delta also teemed with game animals and birds. Elk, deer, antelope, and grizzly bear frequented the tules and the more open countryside. Sightings of elk were reported as late as 1874, but the last of the large game animals are thought to have been destroyed by the 1878 flood.

From the reports of early explorers, it has been estimated that the native population in the Delta area was between 3,000 and 15,000. Most native villages were on natural levees on the edges of the eastern Delta and typically contained around 200 residents, although one community was thought to contain at least 1,000 residents. The native population did not practice agriculture, although they did manage the landscape with fire and other tools to favor plants they used (Anderson, 2005). Their diet consisted of the roots and pollen of the tules, acorns, and the fruit and seeds of other wild plants. Fish and game were also important staples.

European settlement of the Delta began slowly. Despite several expeditions between 1806 and 1812, the Spanish failed to locate a suitable site for missions in the region. From 1813 to 1845, most expeditions were military attempts to subdue the native population. The Hudson Bay Company sent trappers into the Delta from 1828 through 1843 but had limited success because of interference by Native Americans, priests, and local merchants. From 1835 through 1846, the Spanish established several land grants. In 1841, John Sutter was the first foreigner to be granted land in the Delta vicinity. By 1846, an estimated 150 European-Americans were in the Central Valley, mostly at Sutter's Fort near present-day Sacramento. A Dutchman living on an unconfirmed grant below Sutter's Landing was the only certain European-American resident within the Delta, with others scattered on the periphery.

Two events in 1847 set the stage for accelerated settlement of the Delta. The first was the transfer of California to the United States at the end of the Mexican-American war; many U.S. soldiers had volunteered for the war with the idea of staying in California. The second was the introduction of the steamboat, *Sutter's Sitka*. The *Sitka* reduced travel time from Sacramento to San Francisco from a typical two- to three-week trip to just under seven days, a change that greatly facilitated trade throughout the Delta.

Reclamation: Foundations of the Modern Delta Economy

The reclamation of Delta lands began almost simultaneously with the California gold rush. Within weeks of the January 1848 discovery, the few settlements near the coast had all but emptied, and an influx of tens of thousands of people followed. Almost immediately, many miners saw surer fortunes to be made from tilling the soil than from mining. Most of them selected lands on the natural levees of the main waterways or on higher ground near streams close to heavily traveled trails. By the early 1850s, interest turned to the diking and draining of flooded Delta lands.

The reclamation era, which spanned over 80 years, was marked by frequent institutional change, as Delta interests and state and federal authorities sought to tackle problems ranging from basic levee construction, to regional flood control and maintenance of shipping channels, to salinity intrusion. Many of these problems were compounded by the presence of upstream mining activities, which sent massive volumes of debris into the Delta. Although most land reclamation was undertaken by private individuals or local groups, this era witnessed the first major public works project in the Delta—the Central Valley flood control system. By the time the last Delta island was diked and drained in the early 1930s, Delta farmers and the cities on the Delta's periphery had become firmly established interests whose concerns over water quality would figure prominently in the search for large-scale solutions to Delta water issues in subsequent decades.

Reclamation and the Rise of Delta Agriculture

Delta reclamation is a process that becomes increasingly difficult as it progresses. Each acre of drained and diked land represents the removal of floodplains, placing more stress on the remaining system by reducing space for subsequent floodwaters to occupy. Initial reclamation efforts amounted to little more than attempts to supplement natural levees to protect agricultural plots during high tides and seasonal floods. It soon became clear that for reclamation to proceed, institutions were needed to provide land tenure security and to facilitate collective work on levees.

A primary piece of enabling legislation for the reclamation of Delta lands was the Arkansas Act of 1850, more commonly known as the Swampland Act. This law ceded federal swamplands to the states to encourage their reclamation. California received 2,192,506 acres, including nearly 500,000 acres within the Delta. Sales began in 1858. Initially, individual acquisitions were limited to 320 acres, at the price of $1 per acre (about $23 per acre in today's dollars). In 1859, the size limit was doubled to 640 acres, and limits were repealed altogether in 1868.

Although several continuous levees were built in the 1850s (notably, on Grand and Sherman Islands), collective levee building was facilitated by the creation of the Board of Reclamation in 1861, which was given the authority to form reclamation districts from collectives of smaller parcel owners (see Figure 2.1 for the location of individual islands). Between 1861 and 1866, the board authorized reclamation districts to enclose large areas that were defined by natural levees. The board also embarked on several large-scale schemes to reclaim lands and provide flood protection in the Sacramento and Yolo Basins and on several Delta islands. Although the board was dissolved before much of this work could be completed, its duties were transferred to the counties, which continued to oversee the creation of reclamation and levee maintenance districts. Ninety-three of these local agencies still operate within the Delta today, with frontline responsibility for levee maintenance.

Technology also played a central role in reclamation. A contractor in charge of levee construction on Staten Island, J. T. Bailey, developed the first mechanized equipment for levee construction in 1865 (Thompson, 1957). After 1868, when the 640 acre size limit was repealed, corporate speculators and wealthy individuals undertook large-scale reclamation and derived profits from selling the improved land. Machine power was applied to levee construction, land clearing, ditch building, and dredging, and pumps were introduced to drain the parcels.

The influence of these institutional and technological innovations on the pace of reclamation is striking (Table 2.1). In the 1870s, over 90,000 acres were reclaimed, six times more than in the preceding decade.

Table 2.1

Reclamation Growth in the Delta

Decade	Acres Reclaimed	Cumulative Acres
1860–1870	15,000	15,000
1870–1880	92,000	107,000
1880–1890	70,000	177,000
1890–1900	58,000	235,000
1900–1910	88,000	323,000
1910–1920	94,000	417,000
1920–1930	24,000	441,000

SOURCE: Thompson (1957).

Reclamation efforts in the Delta continued through the 1930s, with the last island, McCormack-Williamson Tract, reclaimed in 1934.

In the early years of reclamation, the Delta was seen as a drought-free, fertile area on which the state could depend to support its growth. Delta waterways provided natural and inexpensive transportation routes. The droughts that ruined San Joaquin Valley wheat and barley crops served to further enhance the value of Delta farmlands. An editorial in the San Francisco *Alta* of July 25, 1869, provides a characteristic view:

> In these reclaimable lands we shall have drought-proof means of life and luxurious living for the whole population of our State, were it twice as numerous. Heretofore the certainty of occasional famine years has been a dark cloud on the horizon before the thoughtful vision. Now we see salvation. All hail! to the great minds that have conceived this enterprise. God speed their success and bring them rich reward.

These high hopes waned after the major floods of 1878 and 1881, which revealed the susceptibility of reclaimed lands to recurrent inundations. By this time, however, Delta agriculture had become an important interest in its own right, with landowners seeking relief from floods and mining debris (and, eventually, from salinity intrusion) through judicial and political channels.

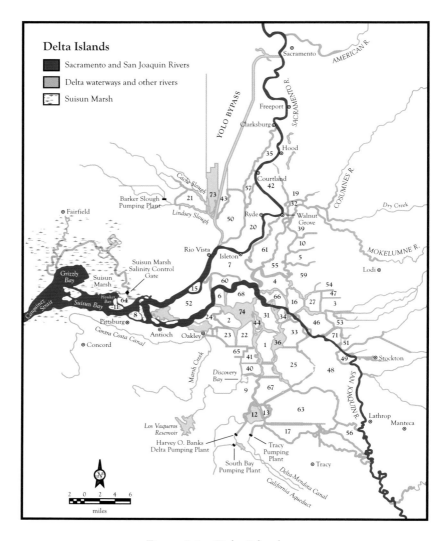

Figure 2.1—Delta Islands

Legal Battles over Upstream Mining

It is estimated that between 1860 and 1914, more than 800 million cubic yards of mining debris—enough to fill 10,000 football fields to a depth of 16 yards—passed through the Delta, primarily from hydraulic mining sites upstream of the Sacramento River watersheds. Although this

Legend for Delta Islands in Figure 2.1

Bacon Island	1	Netherlands	37*
Bethel Tract	2	Neville Island	38*
Bishop Tract	3	New Hope Tract	39
Bouldin Island	4	Orwood Tract	40
Brack Tract	5	Palm Tract	41
Bradford Island	6	Pierson District	42
Brannan-Andrus Island	7	Prospect Island	43
Browns Island	8	Quimby Island	44
Byron Tract	9	Rhode Island	45*
Canal Ranch	10	Rindge Tract	46
Chipps Island	11	Rio Blanco Tract	47
Clifton Court Forebay	12	Roberts Island	48
Coney Island	13	Rough and Ready Island	49
Deadhorse Island	14*	Ryer Island	50
Decker Island	15	Sargent Barnhart Tract	51
Empire Tract	16	Sherman Island	52
Fabian Tract	17	Shima Tract	53
Fay Island	18*	Shin Kee Tract	54
Glanville Tract	19	Staten Island	55
Grand Island	20	Stewart Tract	56
Hastings Tract	21	Sutter Island	57
Holland Tract	22	Sycamore Island	58*
Hotchkiss Tract	23	Terminous Tract	59
Jersey Island	24	Twitchell Island	60
Jones Tract	25	Tyler Island	61
Kimball Island	26*	Union Island	63
King Island	27	Van Sickle Island	64
Little Franks Tract	28*	Veale Tract	65
Little Mandeville Island	29*	Venice Island	66
Little Tinsley Island	30*	Victoria Island	67
Mandeville Island	31	Webb Tract	68
McCormack Williamson Tract	32	Winter Island	69*
McDonald Tract	33	Woodward Island	70
Medford Island	34	Wright-Elmwood Tract	71
Merritt Island	35	Liberty Island	73
Mildred Island	36	Franks Tract	74

NOTE: Numbers with asterisks denote islands not shown on map because of space limits.

debris had some positive effects—notably by bolstering levees and providing fill material—its overall consequences were decidedly negative. The debris raised and constricted the channels, worsening the reduced tidal action caused by reclamation. Consequences included transportation difficulties, increased susceptibility to flooding, and decreased agricultural productivity. (The latter problem, a result of seepage from an elevated water table, was mitigated somewhat when pumps became available in the early 1900s.)

In 1880, the state legislature formed the Board of Drainage Commissioners in an attempt to find a solution between the miners and the farmers. The board was to create drainage basin planning districts with the costs born by a statewide land tax and taxes on hydraulic mining. When this action was invalidated by the State Supreme Court the next year, the farmers instituted injunction proceedings against the miners. The first of these cases—*People v. Gold Run Ditch and Mining Company* (July 1881)—is considered a landmark piece of environmental jurisprudence. It invoked the public trust doctrine to impose an injunction on hydraulic mining. A second case, *Woodruff v. North Bloomfield Gravel Company* (January 1884), also sided with the farmers.

Public Works for Flood Control

In reaction to these rulings and to pressure from Central Valley business interests, subsequent decades saw a flurry of attempts to find a comprehensive solution to flooding issues in the Delta and the greater watersheds of the Sacramento and San Joaquin Rivers. The result was a series of major public investments, involving both the federal and state governments, which are still core elements of the Central Valley flood control system.

The 1893 Caminetti Act authorized the federal government to cooperate with California in formulating plans to prevent mining tailings from passing downstream. The California Debris Commission—a three-member body of Army engineers—was created to work with the federal government in this effort. Although the commission's primary goal was to find a way to resume mining without the tailings problem, its legacy was regional flood control (Kelley, 1989). In 1910, the commission initiated dredging of the lower Sacramento River, under what was known as the

"Minor Project."[2] A commission report submitted to Congress in 1911 formed the basis of a comprehensive flood control plan for the Sacramento River. This plan (dubbed the "Major Project") included proposals for continued channel dredging and the creation of the Yolo Bypass, which provides space for excess water flows on private farmlands.[3] The plan also specified levee heights throughout the Delta.

When California's legislature approved the Major Project in 1911, it also resumed control over reclamation authority, recreating the Board of Reclamation to coordinate state reclamation, flood control, and navigation improvement. The U.S. Congress approved the Major Project in 1917, after the state and landowners agreed to greater participation. The Federal Flood Control Act of 1928 grew from the California Debris Commission's study (as well as Mississippi River experiences) and marked congressional recognition of responsibility in flood control as well as navigation.

Today, flood control within the Central Valley continues to operate under this system of joint responsibility. Federal and state agencies have the primary charge for maintaining roughly 1,600 miles of publicly owned "project levees." Some cost-sharing of project levees is assumed by local reclamation districts and flood control agencies. Within the Delta itself, the mix of responsibilities is more complex. The Delta contains nearly 400 miles of project levees (notably the levees protecting the cities of Lathrop and Stockton) and over 700 miles of "private" agricultural levees, which have limited state cost-sharing (Figure 2.2). Concerns have recently arisen regarding many aspects of the Central Valley flood control system, including the condition of project levees surrounding Sacramento and other upstream locations, but the private Delta levees are a particularly weak link in the system.

---

[2]The Minor Project widened the Sacramento to 3,500 feet and a mean flood stage of 35 feet. Horse Shoe Bend was cut off, Decker Island was created, and a narrow midstream island in front of Rio Vista was removed.

[3]Drawing on the experience with the 1907 flood, the Major Project proposed 600,000 cubic feet per second (cfs) of discharge capability for the Sacramento River. The Yolo Bypass was first proposed in a report by Manson and Grunsky for the Public Works Commission in 1894. Other flood control proposals in this period included that of the Dabney Commission in the early 1900s.

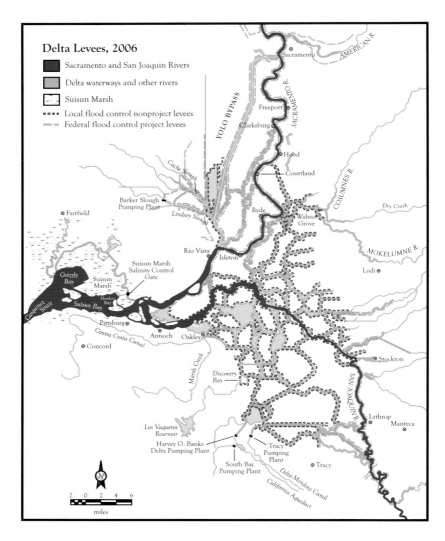

Figure 2.2—Delta Levees, 2006

The Expansion of Shipping Channels

In the early 20th century, the U.S. Army Corps of Engineers also became active in maintaining and improving shipping channels, which had suffered from debris buildup. The earliest efforts focused on the Sacramento corridor. From 1899 to 1927, the corps maintained a channel seven feet deep between Suisun Bay and Sacramento; it was subsequently deepened to 10 feet. In 1946, Congress authorized a project to convert Sacramento into a deepwater port; the dredging of the 30-foot-deep channel was completed in 1955. Similar efforts took place to improve shipping to the eastern Delta. The Stockton channel on the San Joaquin River was maintained at nine feet from 1913 to 1933 and then dredged to 26 feet. In 1950 it was dredged to 30 feet, and in 1987 it was dredged to its current depth of 37 feet at low tide.

These deepwater shipping channels have altered water flows within the Delta.[4] As a result of dredging, water moves much more slowly through the lower Sacramento River than it does in shallower parts of the Delta, thereby providing a different environment for fish and other aquatic life. The Stockton ship channel is particularly important for east-west tidal exchange with the western Delta. Both the Sacramento and the Stockton shipping channels (particularly the Stockton channel) would be threatened by a catastrophic levee failure, which could reintroduce large quantities of sediment into them. At present, these ports are relatively minor players in California's sea trade, although Stockton handles large volumes of agricultural produce from the Central Valley.[5] Sacramento traffic is anticipated to increase under a new management arrangement with the Port of Oakland (Port of Sacramento, 2006).

The First Salinity Lawsuits

By the early 20th century, salinity intrusion had become a major concern for Delta interests. Although it is not certain how far upstream ocean salinity extended under natural conditions, salinity levels did not hamper reclamation in the Delta as they did around the San Francisco Bay (Jackson and Paterson, 1977). In the Delta, virgin reclaimed tracts did not need salts flushed out before agricultural practices began. In this period, salinity intrusion was seasonally highest in the late summer months after the mountain snowpack had melted, and salt water reached farther inland during very dry years, such as 1871 (Young, 1929). However, the

---

[4]The locations of both channels are depicted in Figure 1.2

[5]In 2004, Stockton handled 1.4 percent of total volume and only 0.1 percent of total value of California's sea trade. Sacramento's shares were even lower, at 0.5 percent and 0.06 percent, respectively (www.wisertrade.org)

reduction of tidal floodplains through reclamation and mining debris deposits decreased the penetration of salt into the Delta (Matthew, 1931a). But upstream diversions for irrigation in the Sacramento Valley greatly increased salt intrusion during summer months, especially in dry years. As early as 1908, the sugar refinery at Crockett sent barges as far as 28 miles inland (well into the Delta) to gather fresh water during the dry season (Figure 2.3). During the drought years in the 1920s, salt water reached so far into the Delta that these barges were sent west to Marin instead of east into the Delta. Salt intrusion in the Delta reached its peak between 1910 and 1940, setting the stage for legal proceedings and various engineering proposals to keep the Delta fresh that have continued to this day.

The first salinity lawsuit was filed in July 1920 by the City of Antioch. The city, backed by various Delta interests, charged that upstream irrigators on the Sacramento River were diverting too much water, resulting in insufficient freshwater flows past Antioch to hold back ocean water.[6] Although the lower court initially ruled in Antioch's favor, the California Supreme Court overturned the decision on the basis of evidence showing substantial salinity incursions in the era before significant upstream irrigation.

The suit nevertheless sparked efforts to find engineering solutions to the salinity problem. Initial proposals focused on the construction of a saltwater barrier in the outer part of the estuary, near the Carquinez Strait. A report from the state Department of Public Works (1923) officially endorsed this idea, which had already been considered on several occasions in the second half of the 19th century as a way to control floodwaters and to resolve rail transportation problems across the Delta (Jackson and Paterson, 1977). Further support for a barrier came from those concerned about the effects of an invasive pest, the marine borer *Teredo*, on docks and other wooden structures in the inland ports. This pest, one of the San Francisco Estuary's first invasive species, was moving upstream with salinity incursions. In the end, however, concerns over the high financial costs of a saltwater barrier, as well as the potential harm such a barrier would cause to commercial fisheries, led to its abandonment. Instead, as described below,

---

[6]As discussed in Chapter 6, upstream diversions still have major effects on Delta inflows.

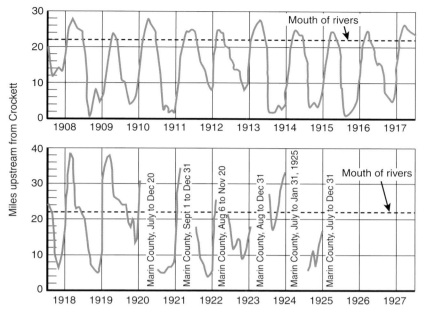

SOURCE: Young (1929), Plate 9-1.

Figure 2.3—Upstream Distance for Barges Looking for Fresh Water for Sugar Refinery at Crockett

control of Delta salinity was woven into projects to augment water supplies for users south of the Delta.

Farming and Land Subsidence

Another problem that increased in severity over time was the subsidence of Delta lands, many of which now lie well below sea level (Figure 2.4). Reclamation itself initiated the subsidence process, because much of the material used to elevate the levees was taken from the interior of reclaimed islands, thereby lowering the island while elevating its protective barrier. Soil burning, mostly associated with the potato farming that developed by 1900, also accounted for much early subsidence. Despite the benefits of burning—weed control, fertilization, and the facilitation of the seedbed—it accelerated subsidence and allowed for salt accumulation and increased wind erosion.

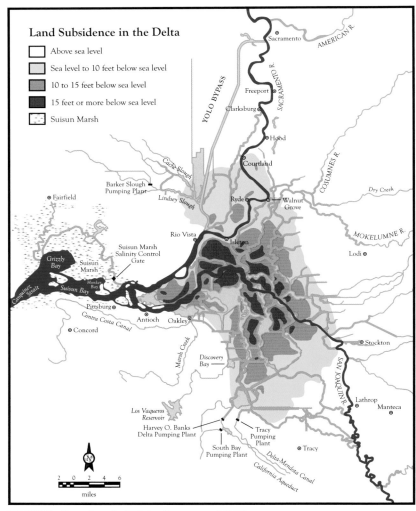

Figure 2.4—Land Subsidence in the Delta

Subsidence added to farming costs because it required additional levee rebuilding, drainage excavation, and pumping both for regular operations and recovery after floods. One casualty of this process was Franks Tract, which was abandoned and left flooded after a 1938 levee failure. The same fate befell Mildred Island in 1983. However, in general, Delta farmers have continued to farm subsided lands. As we will see in Chapter 3, even though the pace of subsidence has slowed in recent times, in part because some of the more destructive farming practices have ceased, subsidence of Delta islands continues and is a major contributor to levee instability.[7]

Big Water Projects Transform the Delta to a Freshwater Body

By the time reclamation of Delta lands was nearly complete in the 1920s, attention began to focus on the development of water supplies from the two major Delta watersheds, the Sacramento and San Joaquin Rivers. Elsewhere in California, major public works projects designed to move water across long distances had already been planned or undertaken, including the Los Angeles Aqueduct (from the Owens Valley to Los Angeles), the Hetch Hetchy project (bringing Sierra Nevada water to San Francisco), the Mokelumne River project (bringing Sierra Nevada water to the East Bay), and the investments along the Colorado River to deliver water to Southern California. From the 1930s to the early 1970s, the Central Valley witnessed a series of major investments in water storage and conveyance to supply agricultural and urban users. This process began with the federally sponsored Central Valley Project (CVP) and ended with the state-run State Water Project (SWP) and included some locally sponsored projects. Although some of the engineering analyses considered alternatives that bypassed the Delta, most of the investments actually undertaken relied on the Delta as a conduit for exports to points south and west (Jackson and Paterson, 1977). As we shall see, big water projects in the Delta have always generated debate, and many plans have been created, modified, and discarded. If nothing else, this process underscores the difficulties of managing the Delta—in the past as well as today.

The Central Valley Project

Since the late 19th century, various observers have recognized the potential for moving surplus Sacramento River water to the drier but

_____

[7]Even in the 1920s, the weakness of Delta levees was seen as a major constraint on Delta solutions, including the design and operation of a saltwater barrier (Young, 1929; Matthew, 1931b).

potentially productive San Joaquin Valley (Alexander, Mendell, and Davidson, 1874). The 1923 Department of Public Works' report to the legislature noted above included proposals to build upstream storage reservoirs to permit such transfers. These plans were fleshed out in the department's 1930 *State Water Plan* ("the Plan"), which would serve as a blueprint for the Central Valley Project (Department of Public Works, 1930). The Plan concluded that upstream storage along the Sacramento River could simultaneously resolve two principal water problems: water shortages in the San Joaquin Valley, where groundwater overdraft—or pumping in excess of natural recharge—had become a serious concern, and salinity intrusion in the Delta, which would be addressed by creating a hydraulic salinity barrier, with controlled releases of water from upstream storage. Ultimately, the Plan rejected the idea of a physical salinity barrier, arguing that its construction could be postponed until the anticipated growth in San Joaquin Valley water demand used up excess reservoir water.[8] Salinity problems in the East Bay would be resolved by piping Delta supplies via a proposed Contra Costa County conduit. Investments along the Colorado River, meanwhile, were seen as the near-term solution to Southern California's additional water needs.

The Central Valley Project was approved by the legislature and the voters in 1933. Seeking to maximize federal financial contributions in the hard economic times of the Depression, the state handed over control of the project to the federal government. Although construction of one of the CVP's primary components, Shasta Dam, got under way by 1938, state and federal agencies did not agree on the final form of diversions for Sacramento River water until the following decade. USBR had proposed a new canal to route the water around the periphery of the Delta between Freeport and the Stockton area. The final outcome, closer to the state's original proposal, was to divert water through the Delta via a small cross-channel just north of Walnut Grove, from which it would travel south to the pumps. The Delta Cross-Channel, constructed by USBR in 1944, still helps to supply

---

[8]In reaching this conclusion, the Plan's authors drew on several studies conducted in the 1920s, including a 1925 study by the U.S. Bureau of Reclamation (USBR), a 1928 privately financed study on the economics of the barrier (the "Means Report"), a 1929 study for the Department of Public Works (Young, 1929), and the report of the joint federal-state commission appointed in 1930 (the Hoover-Young Commission). Among these, the only report to advocate a barrier was the USBR report. See Jackson and Paterson (1977).

the Contra Costa and Delta-Mendota Canals, which entered service in 1948 and 1951, respectively.

The CVP has also been responsible for some major upstream diversions of water from both the Sacramento and San Joaquin Rivers. Following the construction of the Friant Dam (1942) and the Friant-Kern Canal (1948), the CVP began diverting San Joaquin River water to supply irrigators on the east side of the San Joaquin Valley. Subsequent investments on the west side of the Sacramento Valley, notably the Tehama-Colusa Canal (1980), also increased upstream diversions from the Sacramento River.

The CVP was successful in its primary goals: expelling salt water from the Delta by way of controlled releases from Shasta Reservoir and supplying fresh water to irrigators and some urban users in the San Joaquin Valley and areas west of the Delta. The project also provided benefits to power generation and navigation. However, it was less successful in providing additional flood control protection. Levee failures continued to occur in the Delta whenever the surface elevations of water channels exceeded four feet above mean sea level for more than 48 hours. Moreover, the CVP investments in water supply and salinity control were not considered adequate over the long run, given the anticipated growth in demand for water exports. Since the 1940s, a series of investigations have explored longer-term solutions to these issues. Salinity management in the Delta remains a major issue for the CVP.

The State Water Project

In 1960, California voters authorized the first phase of the State Water Project, which aimed to extend water deliveries from northern watersheds to Southern California cities and to farmers in the Tulare Basin that were beyond the reach of the CVP. Although this project ultimately adopted the same basic approach to water exports as the CVP, relying on the Delta as a transfer point, this approach was not a foregone conclusion. Options that surfaced (or resurfaced) included a saltwater barrier, a highly reengineered and simplified Delta, and a peripheral canal. Investigations into the first two options took place in the 1950s. Peripheral canal investigations continued well into the 1970s, as part of the consideration of the SWP's expansion.

The foundation of the State Water Project was laid in the 1950s, through a series of proposals, plans, and legislative actions. In 1953, the state legislature passed the Abshire-Kelly Salinity Control Barrier Act to reexamine the need for a saltwater barrier. The state Division of Water Resources hired a Dutch consultant, Cornelius Biemond, who was Director of Water Supply for Metropolitan Amsterdam. Biemond rejected the idea of a barrier, proposing instead to reduce the Delta's 1,100 miles of levees to a 450-mile system of master levees. This plan included the construction of both a siphon to take Sacramento River water under the San Joaquin River on its way south and a barrier at the confluence of these two rivers.

By 1957, the newly formed Department of Water Resources discarded the concept of a saltwater barrier in favor of a somewhat modified Biemond Plan and recommended it to the governor and legislature as part of the State Water Project (Department of Water Resources, 1957). Under this proposal, water would be transferred through both a trans-Delta system (the Biemond Plan) and an Antioch Crossing Canal, along the Delta's western edge. Three pumping plants in the south Delta near Tracy would pump supplies farther southward. The Biemond Plan would isolate many Delta channels from tidal action, allowing salinity to be controlled with one-third of the available freshwater flow. In 1959, the Water Resources Development Act was passed to pay for the first phase of the SWP; it was approved by the voters in 1960.

Perhaps reflecting the growing political savvy of Delta interests, the SWP ran into greater public acceptance obstacles than the CVP had. As a precondition to the SWP's advancement, the legislature passed the Delta Protection Act of 1959, which established the legal geographical boundaries of the Delta and stipulated that the state-run SWP, in coordination with the federally run CVP, would be required to maintain Delta water quality standards (i.e., sufficiently low salinity to permit farming and other economic uses). However, Delta interests remained concerned about water quality, and in 1961, the State Assembly Interim Committee of Water rejected the Biemond Plan, stating that it was an imposed solution rather than one worked out in consultation with local interests.

While work began on the SWP's main storage and conveyance components—Oroville Dam and the California Aqueduct—deliberations continued on the ultimate solution for moving water from north to south.

An Interagency Delta Committee was formed to examine Delta water problems. As one alternative, USBR revised the peripheral canal proposal from the 1940s.[9] The committee also examined options for keeping the entire Delta fresh, either with a physical barrier at Chipps Island on the Delta's western edge or through the continued use of controlled reservoir releases to maintain a hydraulic saltwater barrier.

In 1964, the committee released its *Proposed Report on Plan of Development, Sacramento–San Joaquin Delta*, again recommending the peripheral canal but with several refinements, including an increase in the volume of diversions from the Sacramento River to supply south-of-Delta users. The report stressed the intangible environmental benefits of the canal and proposed further work to safeguard the water supplies of western counties. In public hearings, only Contra Costa County raised objections to the canal proposal, while environmental groups remained supportive of it.

The peripheral canal was on its way to becoming a reality. By 1966, DWR had officially adopted the canal as a part of the State Water Project and had reached agreements on cost-sharing provisions with USBR. Public meetings were held to gather local input on proposed canal alignments. While waiting for congressional authorization, the new director of DWR placed the project design on hold but continued with right-of-way purchases. In 1969, USBR released its economic feasibility study and recommended that Congress approve the project. Both chambers of the California legislature issued strong endorsements of the canal. Despite its promising start, this version of the peripheral canal never came to be— other forces were at work that changed the course of the debate about the Delta.

Environmental Concerns Change the Course of Delta Policy Debates

The SWP's plans would all change over the following decade, as California, like the nation as a whole, witnessed the rise of environmental concerns. This shift in public attitudes was reflected in new legal and

---

[9]The proposal was launched in the committee's 1963 report, *Report of the Interagency Delta Committee for Delta Planning* (Jackson and Paterson, 1977).

regulatory frameworks for pollution control and species protection. The Delta and its tributary watersheds, home to many unique aquatic species, would become a focal point for these new concerns. One casualty would be the build-out of the State Water Project, as northern rivers slated as sources for additional upstream storage were declared "Wild and Scenic" and off limits for new reservoirs or diversions. Another casualty would be the peripheral canal, which eventually drew strong environmental opposition.

The wave of new environmental legislation began in the mid-1960s, with a succession of federal laws regarding water quality and species protection—the National Wilderness Preservation Act (1964), the Federal Endangered Species Preservation Act (1966, a precursor to the 1973 Endangered Species Act), the National Wild and Scenic Rivers Act (1968), the National Environmental Policy Act (1969), the Clean Water Act (1972), and the Safe Drinking Water Act (1974). California's legislature was equally active in the environmental arena, passing comparable bills at the state level.

As species protection became an explicit goal in the Delta, alongside the maintenance of fresh water for human uses, perceptions of the effects of water diversions and the nature of water quality problems began to change. In 1971, the State Water Resources Control Board (SWRCB) adopted Water Rights Decision 1379, establishing water quality standards for the CVP and the SWP that included new outflow requirements for the San Francisco Bay–Delta Estuary and a comprehensive monitoring program to follow changes in environmental conditions. This decision, stayed by court order in response to lawsuits filed by San Joaquin Valley irrigation districts, marked the beginning of a series of legal and regulatory battles over Delta water quality standards for the environment.[10]

---

[10]In 1978, the SWRCB adopted a new water quality control plan for the Delta and Suisun Marsh (the 1978 Delta Plan) and set new Delta water quality standards with Decision 1485 (D-1485), again focusing on environmental as well as human water quality needs and implying greater restrictions on water exports. Following successful legal challenges at the trial court level, the 1986 "Racanelli Decision" affirmed the SWRCB's broad authority and discretion over water rights and quality issues, including jurisdiction over the CVP. The SWRCB was ordered to prepare a new plan for Delta flows and export guidelines with a greater environmental emphasis. This new draft, put forth in 1988, was withdrawn the following year amid controversy over its legal and water rights implications.

Defeat of the Peripheral Canal

During the 1970s, the peripheral canal plan was also subject to increased environmental scrutiny. Although the canal was initially promoted as having environmental benefits in addition to the primary benefit of controlling the salinity of Delta water exports, these benefits were not spelled out in any detail in the reports of the 1960s. Subsequent reports were more mixed. Controversy around the plan began to build, generating considerable debate, including lawsuits, over several years.[11] In the end, the canal was beaten in the court of public opinion. By the time it was put to a referendum in 1982, an alliance of environmentalists and northern water interests, with backing from some Tulare Basin farmers who feared water high costs (Arax and Wartzman, 2005), successfully argued that the canal would be bad for the environment and Northern California water rights. Large majorities of Northern California voters rejected the perceived water grab by Southern California.[12]

Drought Intensifies Conflict

In 1987, California entered a multiyear drought that severely reduced available flows from the Delta's two main watersheds. As the drought wore on, it provoked conflict over the amount of water reserved for environmental flows. Initially, CVP and SWP exports were not cut, and both environmentalists and fisheries agencies raised concerns over the consequences for important fish species that depended on the Delta. In 1989, the Sacramento River winter-run Chinook salmon was listed as

---

[11] In 1970, a preliminary report from the U.S. Geological Survey suggested that the southern San Francisco Bay could suffer from reduced Delta outflows. A 1973 report by the director of the California Department of Fish and Game endorsed the canal for correcting adverse conditions in the Delta for fish (notably problems caused by pumping in the southern Delta), but it also stressed the importance of maintaining adequate flows within the Delta itself and of involving fisheries agencies in the decisionmaking process (Arnett, 1973). That same year, a student uncovered an unknown, preliminary report from the federal Environmental Protection Agency (U.S. EPA) that was highly critical of the canal. The student gave the report to the Friends of the Earth and it was made public. DWR published a 600-page draft *Environmental Impact Report* in August 1974 with only minor changes from the 1969 design. In the early 1970s, environmental groups filed a series of complaints and lawsuits on a range of procedural issues relating to federal involvement and permitting of the peripheral canal (Jackson and Paterson, 1977; Hundley, 2001).

[12] In Northern California counties, the "no" vote consistently exceeded 90 percent. Strong majorities in all San Joaquin Valley counties except Kern also rejected the canal.

threatened under the federal Endangered Species Act and as endangered under its state counterpart, and DWR and USBR agreed to build salinity control gates in Suisun Marsh and make other efforts to preserve the habitat in the marsh.

With the drought still in full force, water exports to some San Joaquin Valley farmers were reduced in 1991 to maintain minimum environmental flows. The following year, water users were dealt several legal and legislative blows.[13] By 1993, a crisis was erupting. The delta smelt was listed as a threatened species, and other listings began to follow (Table 2.2). The federal EPA threatened to impose stricter water quality standards for the estuary that would severely curtail water exports. Under the threat of a regulatory hammer, water users agreed to work with environmental interests to forge a new plan for the Delta that would comprehensively address both water user and environmental concerns. In December 1994, the signing of the Bay-Delta Accord marked the beginning of the CALFED era.

The CALFED Era: Testing the Limits of Consensus

CALFED sought to involve the full array of relevant federal and state agencies, together with local and statewide stakeholders, to form a new plan for the Bay-Delta. The CALFED process continued in earnest for roughly a decade, funded primarily with state bond monies and some limited federal contributions.

One of CALFED's early efforts was to review and compare strategic alternatives for the Delta. Over 20 diverse conceptual alternatives were initially reviewed and briefly discussed, but little formal analysis was published (CALFED, 1996). The CALFED Record of Decision (ROD) was signed in mid-2000 by all agencies with authority over Delta operations, and it advocated the continuation of the through-Delta strategy for water exports. All four of CALFED's main goals (water supply

---

[13]The courts upheld that an irrigation district must cease pumping during peak migration times for endangered Chinook salmon and that the CVP must release flows sufficient to protect downstream fisheries. Congress then passed the Central Valley Project Improvement Act (CVPIA), a central component of which was a requirement that the CVP commit 800,000 acre-feet/year (or roughly 10 percent of total deliveries) to support fish and wildlife.

Table 2.2

Status of Fish Species in the Sacramento–San Joaquin Delta Watersheds

Species	Year	Status
Sacramento River winter-run Chinook salmon	1989	Endangered (CESA) Threatened (ESA)
Delta smelt	1993	Threatened (ESA and CESA)
Sacramento River winter-run Chinook salmon	1994	Reclassified as endangered (ESA)
Sacramento splittail	1995	Species of concern (CESA)[a]
Longfin smelt	1995	Species of concern (CESA)
Sacramento perch	1995	Species of concern (CESA)
River lamprey	1995	Species of concern (CESA)
Central Valley steelhead trout	1998	Threatened (ESA)
Central Valley spring-run Chinook salmon	1999	Threatened (ESA)
Sacramento River drainage spring-run Chinook salmon	1999	Threatened (CESA)
Central Valley fall-run and late-fall-run Chinook salmon	2004	Species of concern (ESA)
Southern green sturgeon	2006	Threatened (ESA)

SOURCE: Department of Fish and Game (2006a), available at www.dfg.ca.gov/hcpb/species/t_e_spp/tefisha/tefisha.shtml.

NOTES: ESA and CESA refer to the federal and California Endangered Species Acts, respectively.

[a]The Sacramento splittail was listed as threatened under the ESA in 1999 but was removed from the list in 2003.

reliability, water quality, ecosystem restoration, levees) were based on this strategy and were not to be revisited until 2007. The maxim that "everyone would get better together" tied all fates to this single approach.

CALFED proved to be a fragile truce. As discussed in more detail in Chapter 5, by the tenth anniversary of the Bay-Delta Accord, stakeholder frustrations were widespread. Water exporters were frustrated with slow movement to augment water supplies, which in some cases meant restoring supplies that had been reduced to support the environment. In-Delta users were discouraged by the limited progress on dealing with Delta salinity and water quality. Environmental interests remained concerned that water export goals were taking precedence over ecosystem protection—a concern that turned into alarm when the news broke about precipitous drops in the delta smelt and other pelagic fish species. And Delta landowners and farmers were frustrated over limited funds for levee improvements and maintenance, which had previously received some state funding but were not a priority for CALFED funds.

Arguably, CALFED was not designed to deal with some of the problems that have recently emerged. New research on the long-term risks associated with Delta levees, the significant levee breach on Jones Tract in the summer of 2004, and the devastating effects of levee breaches in New Orleans all made the levee issue more urgent than it had been in the years leading up to the CALFED ROD. Similarly, CALFED's initial ecosystem focus was on restoring salmon runs, in part because delta smelt and other pelagic organisms were less understood. The recent severe declines in these fish populations caught most experts by surprise.

CALFED was also founded on the implicit assumption that the Delta would not face the urbanization pressures that have become apparent over the past few years. This assumption may have been justified in the early to mid-1990s, particularly in light of the passage of the Delta Protection Act of 1992, which reserved most Delta lowlands for agricultural and environmental uses. However, since the late 1990s, a housing boom has swept the Central Valley, and today a number of large projects are slated for development in lowland areas that are exempt from the act's restrictions. In addition, recent concerns about urban flood risks behind agricultural levees, state liability for failure of project levees (following the 2003 *Paterno* decision), and the long-term environmental effects of urbanizing Delta islands have raised urbanization as a serious long-term issue for Delta management.[14]

But CALFED also suffered from some fundamental design flaws, particularly with regard to financing. CALFED parties agreed to a principle of "beneficiary pays," but in practice, the implications for user contributions were never fleshed out. The program was launched at the height of the dot-com boom, when the state enjoyed windfall surplus revenues, and it relied on unrealistic expectations of massive state and federal taxpayer funds. Serious, long-term funding proposals were never developed. This lack did not matter so much in the first years after

---

[14]For more on *Paterno*, see Department of Water Resources (2005a).

the signing of the ROD, because $1.5 billion in state bond funds was earmarked for the program (de Alth and Rueben, 2005). But by 2005, when most bond funds had run out, legislative frustration over the lack of a realistic plan for beneficiary contributions spelled the end of most CALFED activities.

CALFED did achieve some notable successes. Major improvements were achieved in interagency coordination. Considerable progress was made in ecosystem restoration in several watersheds upstream of the Delta. Water transfers have become largely accepted statewide, with success during the 1987–1992 drought followed by a very successful Environmental Water Account (Hanak, 2003). Improvements in water conservation efforts have continued, and funding for research has brought more data and some new thinking to Delta ecological problems. Ultimately, however, the program suffered from a failure of political processes to come to long-term agreement without continued massive taxpayer subsidies. In light of the new problems facing the Delta, it now appears that the CALFED premise that everyone can get better together may be unrealistic.

The Lessons of Delta History

The Delta's short history of European settlement has seen major changes in the form, use, and settlement of land in the Delta. Before European settlement, the Delta was a massive tidal marsh, with significant seasonal variations in flow and salinity, as well as large interannual variations caused by floods and droughts. This era was followed by a period of land reclamation for agriculture, which, for better or worse, created much of the Delta's current landscape. Marsh reclamation reduced tidal flows, but upstream diversions in the Sacramento Valley increased salinity intrusion into the central Delta during dry seasons of dry years, processes clearly understood in the 1930s.

The prospect of major water exports from the Delta made salinity intrusion a primary concern for all water users within the Delta. Various strategies, including saltwater barriers, were considered early on. By the 1930s, a hydraulic barrier, consisting of Delta outflows from upstream reservoirs, was selected as the primary means of salinity control for agricultural and urban water users. Using this approach, both in-Delta users and water exporters could agree on a need to keep the Delta always fresh.

The notion of an always-fresh Delta supported by persistent net Delta outflows has endured for over 70 years, but it is not aging well. This management strategy retains support from in-Delta users, but water exporters have come to see increasing risks from this approach, for reasons described in Chapter 3. In Chapter 4, we will examine changes in our understanding of the Delta ecosystem, which also cause us to doubt the wisdom of continuing with this strategy. Because of the history of profound and widespread change in the Delta, we are long past the point where the Delta can be "restored" to past conditions, whether it be the pre-European Delta or the bucolic agricultural Delta. No matter what we do, the Delta of the near future will be very different from past Deltas.

Delta history provides insight into the processes by which Californians have sought solutions to collective problems in this pivotal region. And as this history suggests, these processes have rarely been simple or smooth. At several points over the last century, strenuous efforts have been made to provide solutions to the Delta's problems, and these solutions have been followed by major investments in the chosen strategy. From the 1890s to the 1910s, the Debris Commission worked on Central Valley flood control. Later, state and federal efforts developed the 1930 State Water Plan and executed the Central Valley Project; investigations in the 1950s led to the development of the State Water Project. In more recent times, as environmental concerns have become central in Delta policy considerations, the search for solutions appears more constrained. Thus, CALFED worked under the premise that the Delta's basic configuration should remain unchanged and that environmental goals could be satisfied simultaneously with those of exporters and in-Delta interests. Given the crisis now looming in the Delta, it is once again time for California to launch a serious search for solutions, both old and new.

3. Drivers of Change Within the Delta

"... danger is never so near as when you are unprepared for it."

Francis Parkman (1849), The Oregon Trail

As we have seen in the last two chapters, the Delta has provided an array of services to the people and economy of California for the past 150 years. These diverse services—ranging from water supply to farming to shipping to recreation—have all required some manipulation of the hydrology and the landscape of the Delta. The construction of dikes and the draining of marshlands to support farming is the most regionally significant and visible physical manipulation. Maintaining water quality standards to sustain exports, in-Delta water diversions, and ecosystem needs has required sophisticated hydrologic and landscape engineering. Even low-profile services, such as hunting, fishing, and boating, require significant maintenance interventions.

The development of the Delta has completely transformed the region, leaving no significant remnants of the original landscape (Bay Institute, 1998). This transformation has been both dramatic and, on a geological time scale, instantaneous. When framed within the overall changes in California since the gold rush, the scope and scale of the Delta's transformation is on par with other rapid changes throughout the state, particularly within the major urban centers and the agricultural valleys. The Delta, like many other regions of California, exhibits a complex mix of natural responses to human-induced changes and has experienced numerous unintended and often undesirable consequences. If present trends continue, several uncontrolled hydrologic, ecologic, and landscape changes will occur into the indefinite future and pose great threats to the sustained provision of Delta services. Unfortunately, these changes appear to be outpacing the abilities of both the scientific community and policymakers to keep up.

All naturally evolving landscapes undergo a process of constant feedback between landscape processes and such drivers of landscape change as tectonic activity (changes resulting from movements in the Earth's crust), sea level change, and climate change. This process is particularly pronounced in estuarine, coastal, riverine, and deltaic systems, in which subtle changes in certain landscape drivers, including runoff, sediment supply, and tide and wave energy, are accommodated by corresponding changes in patterns of deposition, erosion, and landscape form (Pethick and Crook, 2000; Reed, 2002). In theory, this kind of feedback maintains a dynamic equilibrium, in which the landscape is in rough balance with the forces acting on it, even as it changes over the long term. In practice, because of human activity, the Delta is in profound and increasing disequilibrium with the forces currently operating on it.

This chapter outlines several key drivers of change within the Delta. The focus here is on natural and human-driven changes that not only affect our ability to benefit from Delta services but are also likely to significantly reduce the quality of these services in the future. The six key drivers, discussed in a recent CALFED report by Mount, Twiss, and Adams (2006), include land subsidence, sea level rise, seismicity, regional climate change, alien species, and urbanization.

Subsidence and Sea Level Rise

The most significant and enduring effect on Delta landscapes has been the conversion of roughly 450,000 acres of freshwater tidal marsh into farmland during the late 1800s and early 1900s. The draining and tilling of the Delta's organic-rich soils initiated a period of subsidence, a rapid lowering of land surface elevations of Delta islands perhaps unmatched in the world. The location and magnitude of subsidence has been and will continue to be the greatest influence on the Delta's landscape and is a fundamental constraint on future efforts to manage the Delta's services.

The exceptional subsidence of the Delta stems from its unique geologic setting and historical land use practices. For more than 6,000 years, the Delta was a freshwater tidal marsh (Shlemon and Begg, 1975; Atwater, 1982) consisting of a complex network of tidal channels, sloughs, "islands" composed of tule marsh plains, complex branching ("dendritic") water channels, and natural levees colonized by riparian forests (Bay Institute, 1998). A slow rise in sea level and gradual regional tectonic subsidence (subsidence of the land resulting from flexure of the Earth's crust) created what geologists refer to as "accommodation space" and made room for the relatively continuous accumulation of large volumes of sediment within the Delta (Atwater et al., 1979; Orr, Crooks, and Williams, 2003). Analysis of core samples by Shlemon and Begg (1975) and Atwater (1982) suggests that as accommodation space was formed by sea level rise over the last 6,000 years, it was quickly filled by the deposition of inorganic sediment from the Sacramento and San Joaquin Rivers and a similar amount of *in situ* production of organic material in the tule marshes. The preservation of this material, as the peat soils of the Delta, benefited from the oxygen-poor conditions within saturated soils of the marshes.

These natural patterns were substantially altered by reclamation in the late 1800s and early 1900s (Mount and Twiss, 2005). As we saw in Chapter 2, to farm the organic-rich soils, farmers needed to drain the islands. This involved constructing levees around the islands, filling most tidal channels and sloughs, and, most important, lowering local groundwater tables below crop root zones by constructing perimeter drains.[1] The draining of Delta soils caused widespread elevation loss.[2] This process was exacerbated by destructive land use practices, including peat burning and tillage, which promoted wind erosion (the most destructive practices are no longer used). The pace of subsidence was exceptional, exceeding four inches per year on some islands with the most intensive practices. Today, all islands of the Delta that contained peat soils and were used for agriculture have subsided; most in the central and western Delta lie more than 10 feet below today's mean sea level (Figure 3.1).[3]

Modeling Subsidence

The rapid loss of island elevation during the 20th century created a new form of human-induced or "anthropogenic" accommodation space below sea level. This space has no natural analog. It has not filled with either sediment or water, as would occur normally in an estuary capable of natural

---

[1] Such drainage systems prevent waterlogging of a property—in this case, the Delta island. For an illustration, see Figure 3.1.

[2] See Deverel, Wang, and Rojstaczer (1998) and Deverel and Rojstaczer (1996). Contributing factors included microbial oxidation of organic matter, consolidation as a result of dewatering, and compaction of underlying soils.

[3] For a map of subsidence levels in the Delta, see Figure 2.4.

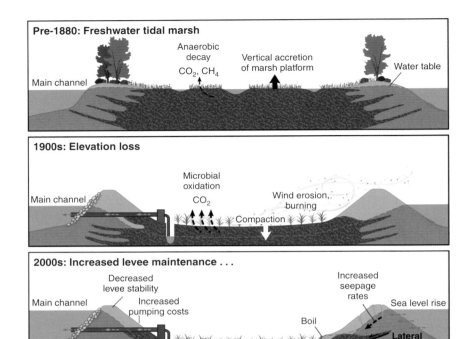

Pre-1880: Freshwater tidal marsh

Anaerobic decay CO_2, CH_4

Vertical accretion of marsh platform

Water table

Main channel

1900s: Elevation loss

Microbial oxidation CO_2

Wind erosion, burning

Main channel

Compaction

2000s: Increased levee maintenance . . .

Decreased levee stability

Increased pumping costs

Main channel

Boil

Increased seepage rates

Sea level rise

Lateral deformation

. . . Or levee failure

Figure 3.1—Conceptual Diagram Illustrating the Historical and Future Trajectory of Island Subsidence in the Delta

adaptation but is instead filled with air (as shown in the second and third panels of Figure 3.1).

Using a simplified geographic model, Mount and Twiss (2005) tracked the formation of this accommodation space in the Delta over the past 100 years. Their results indicate that more than 3.4 billion cubic yards of space has been created, roughly equivalent to 70,000 football fields 30 feet deep, or the volume of material used to construct Rome (Hooke, 2000). Mount and Twiss then used the same model to project future subsidence in the Delta over the course of the next 50 years. This model assumed that the Delta would continue to be farmed and that peat oxidation would continue to generate accommodation space. It also factored in sea level rise over the next 50 years, which magnifies the effect of subsidence by increasing the differential between interior island elevations and water surface elevations.[4] The results, summarized in Figure 3.2, suggest that under business-as-usual conditions, the Delta will generate an additional 1.3 billion cubic yards of accommodation space. However, the patterns of subsidence will change during this time. In the southern Delta and portions of the eastern Delta, where farming practices have completely removed the peat soils, sea level rise is the only driver of new accommodation space. But in the central, western, and northern Delta, if the lands continue to be farmed, subsidence will continue for much of the next century—in other words, agriculture will also drive the creation of accommodation space (Figure 3.3).

Subsidence, Sea Level Rise, and Levee Failure

The creation of accommodation space by human activity has the unintended effect of putting the landscape in considerable disequilibrium. Water is seeking to refill the subsided islands. This state of imbalance is maintained by more than 1,100 miles of artificial levees (Department of Water Resources et al., 2002), which are increasingly subject to failure. Levee failure and subsequent island flooding can have many causes (including such mundane things as burrowing by beavers and ground squirrels), some of which have no direct relationship to the magnitude of land subsidence. However, on a regional and local scale, the difference between interior island elevation and adjacent channel water surface elevation is a useful measure of the relative magnitude of the forces acting on levees. The greater these forces, the greater the potential for water seepage through and under levees—a common cause of levee failure.

Mount and Twiss (2005) developed a simplified measure of levee failure potential in the Delta as a function of island subsidence and sea level

---

[4]Conservative estimates of sea level rise were factored into the model using values provided by the Intergovernmental Panel on Climate Change (2001).

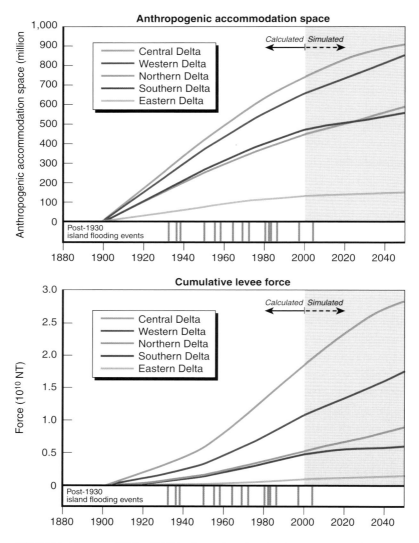

SOURCE: Mount and Twiss (2005). Reproduced under the terms of the Creative Commons Attribution License.

Figure 3.2—Historical and Projected Changes in Anthropogenic Accommodation Space and Cumulative Hydrostatic Force in the Delta

rise over the next 50 years. They calculated the hydrostatic forces (that is, the pressure exerted by water) acting on levees throughout the Delta; these

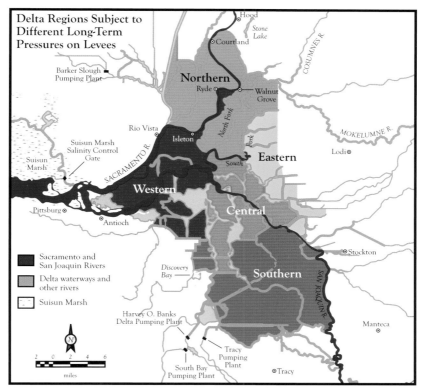

NOTE: The regional designations shown here were used to calculate pressures on the levees, as depicted in Figure 3.2. They often differ from areas of the Delta discussed elsewhere in this report. For instance, the "western" region shown here extends farther to the north and east than the area considered for fluctuating Delta salinity in subsequent chapters.

Figure 3.3—Delta Regions Subject to Different Long-Term Pressures on Levees

forces increase with the squared difference between land and water heights. For each island, they estimated total hydrostatic force over the island's entire levee length. Using this approach, they found that deeply subsided islands have a high cumulative hydrostatic force and thus a high potential for failure. Islands with long levee lengths also have a high potential for failure because of the greater opportunity for hydrostatic pressure to exploit local levee weaknesses. Deeply subsided islands with long levee lengths are at the highest risk of future failure. Figure 3.2 depicts the historical and projected changes in cumulative hydrostatic force. These estimates indicate that the central and western Delta, in particular, will be increasingly vulnerable to levee failures and island flooding over the next 50 years and into the indefinite future.

Levee Policy

Although the Central Valley flood control system established in the 1910s set minimum heights for Delta levees, state regulatory involvement in the many privately owned levees remained negligible for most of the 20th century. Following the large 1986 flood in the Central Valley, which exposed the poor condition of Delta levees, the state legislature established new levee standards and launched a program of financial support. Supported by Senate Bill 34, the Delta Levee Subventions Program provided funds to maintain and upgrade levees, with the goal of raising levee crowns to one foot above the estimated 100-year flood stage height to meet State Hazard Mitigation Plan standards (Department of Water Resources, 1995). A long-term goal for the Delta is to meet Federal Public Law (PL) 84-99 standards for agricultural levees.

The subventions program, which dedicated roughly $110 million in state funds and $90 million in local matching funds to Delta levees between 1988 and 2005, has noticeably improved the conditions of many levees. However, it is important to recognize the program's limitations. Upgrading levees to meet the program's target elevation does not guarantee that Delta levees will not fail during a 100-year flood event (100-year floods have a probability of 1 percent of occurring in any given year). The one-foot difference between the estimated 100-year flood stage height and the levee crowns, particularly in a region subject to very high winds during floods, is insufficient to prevent levee failure. Moreover, the subvention program did not address the interior or the foundation of most levees, so seepage under and through levees remains an important threat during high water flows and could cause levees to fail even before they are overrun by floodwaters. Finally, the 100-year standard elevation estimate was based on 1986 hydrology rather than current hydrology, which takes into account changes in runoff conditions (discussed below). The National Flood Insurance Program maps, which have not been updated recently, place the entire Delta into the 100-year floodplain, reflecting the relatively low level of protection that the levees provide. It is reasonable to assume that in the future, large inflows of water into the Delta will inevitably result in multiple island failures.

Seismicity

For more than 30 years, DWR has warned that earthquakes pose considerable risk to Delta levees (Department of Water Resources, 1995). At least five major faults lie within close proximity to the Delta and are capable of producing significant ground accelerations. Poor foundation soils and poor-quality levee construction materials lead to a high risk of failure caused by liquefaction and settling.[5] Multiple seismic risk studies conducted for the Bay Area indicate a very high potential for major quakes in the region in the near future.[6]

In a report prepared for the CALFED Levee System Integrity Program, Torres et al. (2000) showed that ground accelerations from moderate earthquakes (magnitude 6.0, with a probability of recurring on average every 100 years) are capable of causing multiple levee failures. The highest risk of levee failure is in the western Delta, because of deep subsidence, poor foundations, and proximity to several significant seismic sources. However, a medium to high risk of catastrophic levee failures exists for almost all the central Delta as well.

Some local Delta engineers judge that seismicity is not a problem for the Delta because no local levee collapses have occurred from earthquakes in the past. However, there have been no significant ground accelerations in the Delta since the 1906 earthquake, before tall levees were constructed to protect subsided islands. The levees that now protect deeply subsided islands have not yet been tested. Moreover, the State Hazard Mitigation Plan and federal PL 84-99 standards do not address the susceptibility of levees and their foundations to failure during seismic shaking. Upgrading levees to meet PL 84-99 standards—at an estimated cost of roughly

---

[5]Liquefaction is the tendency of some soils to behave like a liquid when shaken, as happened in the Marina District of San Francisco during the 1989 earthquake.

[6]See http//quake.usgs.gov/research/seismology/wg02/.

$1 billion to $2 billion—will do little to reduce the potential for failure during earthquakes.

Seismicity poses a significant threat to the management and maintenance of current and future services provided by the Delta. Preliminary consequences of a rare, large quake would likely be that 16 or more islands would flood, principally within the central and western Delta (Jack R. Benjamin and Associates, 2005). All modeling to date indicates that this flooding would significantly alter the volume of the tidal prism (i.e., the volume of water moved during each tidal cycle) and local hydrodynamics with severe, prolonged disruptions in water quality and aquatic habitat.

The risk of sudden change in the Delta is quite high. In a simplified review of this risk, Mount and Twiss (2005) evaluated the probability of a major event that would significantly and perhaps permanently change the configuration of the Delta abruptly. Their analysis highlighted two sources of potential dramatic change: major seismic events and floods that are likely to recur every 100 years or less. Their calculations show that the probability is roughly two-in-three that during the next 50 years either a large flood or seismic event will affect the Delta. However, this analysis underestimates the actual probabilities for two reasons. First, strain continues to accumulate on Bay Area faults, increasing the annual risk of seismic activity. Second, current calculations of the size of a 100-year flood in the Delta are based on outdated hydrology data, which neglect the much higher inflows from rivers feeding into the Delta in recent years. In sum, the Delta is likely to change significantly and abruptly during the next generation. Sudden catastrophic change would be a very hard landing indeed for those depending on the Delta.

Regional Climate Change

Approximately 50 percent of California's average annual runoff, derived from roughly 45 percent of its surface area, flows to the Sacramento–San Joaquin Delta. The magnitude, timing, and duration of these inflows are, along with tides, the major influence on the physical and biological conditions that dictate the services that can be derived from the Delta. Regional climate change, driven principally by the Earth's warming in response to increases in greenhouse gasses, is currently affecting inflows to the Delta and will continue to affect them into the indefinite future (Knowles and Cayan, 2004; Hayhoe et al., 2004; Department of Water Resources, 2006).

Since the latter half of the 20th century, there has been a general trend toward increasing hydrologic variability and changes in the timing of runoff in the western United States (Jain, Hoerling, and Eischeid, 2005; Stewart, Cayan, and Dettinger, 2004). This trend has been particularly pronounced for the Sierra Nevada mountains and the Central Valley (Aguado et al., 1992). The region also has witnessed increased frequency and intensity of extreme rainfall events. Additionally, there has been a long-term shift in the seasonal pattern of runoff, with peaks shifting from spring toward winter (Dettinger et al., 2004). These changes in runoff are consistent with the results of regional climate models.[7]

Most modeling efforts predict that in the coming century, California will see a continuation of the hydrologic and climatologic trends established in the latter half of the 20th century (Dettinger, 2005). Warming trends will continue, with an increase in average annual temperatures of 2°F to 5°F by the 2030s and 4°F to 18°F by 2100 (Hayhoe et al., 2004). Recent work suggests significant increased interannual variability (vanRheenan et al., 2004) with the potential for increased frequency of both critically dry and wet years (Maurer, 2006) and significant declines in summer and fall inflows to the Delta because of shifts in the timing of snowmelt runoff (Zhu, Jenkins, and Lund, 2005; Miller, Bashford, and Strem, 2003). Additionally, regional models generally depict significant increases in the number of large winter storms, with associated increases in high winter inflows to the Delta.

The effects of ongoing and future changes in climate and runoff on the Delta have not been well documented to date, but they are the subject of numerous research efforts.[8] Water resource and flood management operations will be able to mute many of the effects of climate change, with the possible exception of increases in water temperature associated with increases in ambient air temperatures (Tanaka et al., 2006). However,

---

[7]To derive predictions for individual regions such as California, global climate models, known as General Circulation Models (GCM), are "downscaled."

[8]For a summary, see Department of Water Resources (2006).

all changes point toward a long-term, multidecade decline in the quality of Delta services. First, the increased frequency and magnitude of winter floods in the Delta will exacerbate pressures on the levee network, raising the cost of maintenance and increasing the likelihood of widespread, multi-island floods. In principle, reservoir operations can be altered to reduce the peak flood flows. In practice, however, there is likely to be growing conflict between flood control and water supply goals for reservoir management. To make sure that they store enough water for summer use, managers will face pressure to fill reservoirs during the winter rather than during the spring when runoff is likely to be less reliable. Yet such a strategy might increase flood risks, given the growing likelihood and magnitude of winter floods. Second, climate change is likely to introduce significant water quality costs. Currently, during low inflow periods, water quality in the Delta is generally poor, owing to the poor water quality of the San Joaquin River and to salinity intrusions from the Bay, coupled with increases in the influence of tides. Over the course of the next century, the shift in timing of runoff from spring to winter and the increase in frequency of critically dry years suggest long-term declines in Delta water quality, with a wide range of effects.

Alien Species

The San Francisco Bay–Delta is arguably one of the most invaded estuaries in the world (Cohen and Carlton, 1998). More than 250 alien species of aquatic and terrestrial plants and animals have entered the estuary since the first arrival of Europeans, with most indications showing that the pace of invasions has increased in recent decades (Figure 3.4). At least 185 alien species now inhabit the Delta and have profoundly changed Bay-Delta food webs and habitats, generating an array of effects—mostly negative—on native species. They also contribute to levee problems (e.g., burrowing by muskrats and crayfish), impede navigation (e.g., floating mats of water hyacinth), and otherwise cause economic damage. Today and for the indefinite future, we are managing an ecosystem composed of a mix of native and alien species that are in constant flux, as native species decline in abundance, new alien species invade, and established aliens wax and wane in numbers.

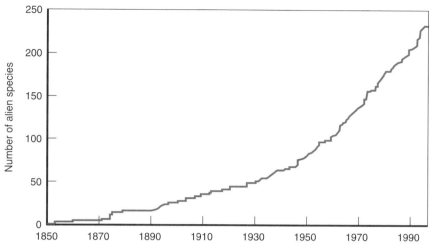

SOURCE: Cohen and Carlton (1998).

Figure 3.4—Estimated Number of Alien Species Within the San Francisco Estuary, 1850–1990

Although we have an improved ability to predict the effects of species invasions (e.g., Moyle and Marchetti, 2006), the process of invasion remains highly idiosyncratic in terms of which aliens will be most successful and change the ecosystem they invade. Nevertheless, several alien species not yet established in the Delta, such as the zebra mussel, are likely both to invade and to have large effects (Table 3.1). Invasions of alien species continue because efforts to halt new invasions have been small compared to the magnitude of the problem (e.g., Nobriga et al., 2005). For this reason, invasions by alien species and changes in the abundance of established alien species are another driver of change in the Delta. (Chapter 4 discusses this issue in greater depth.)

Urbanization

Although population growth has slowed in California in recent decades, the absolute population increases anticipated over the coming decades remain dramatic. By 2025, the state is expected to add another nine million residents—more than the population of the state of Ohio—

Table 3.1

Examples of Alien Species That May Invade the Delta in the Near Future

Species	Threat Rating	Invasion Likelihood	Likely Source	Why a Threat	Comments
Fish					
Northern pike	1	1	Lake Davis, California	Predator on salmon, native fish	Eradication program scheduled
White bass	2	1	California reservoirs	Schooling predator on pelagic fish	
Grass carp	2	2	Illegal import from Southern California	Changes aquatic communities	Used for aquatic weed control
Silver carp	2	3	Illegal import from Eastern United States	Plankton feeder	One of three Asian carp species infesting Eastern United States
Invertebrates					
Zebra mussel	1	2	Recreational boats, ballast water	Changes food webs, clogs water supply systems	Spreading rapidly in United States
Spiny waterflea	2	3	Ballast water	Predator on zooplankton	Two species, one fresh, one marine
Atlantic comb-jelly	1	2	Ballast water	Predator on zooplankton	Caused major changes to Black Sea ecosystem
Asian fish tapeworm	2	1	Imported live fish	Kills, weakens native fish	Example of disease/parasite

Table 3.1 (continued)

Species	Threat Rating	Invasion Likelihood	Likely Source	Why a Threat	Comments
Plants					
Oxygen weed	2	2	Aquarium trade	Clogs waterways	Aquatic plant
Water chestnut	2	1	Aquarium and ornamental trade	Forms dense mats	Aquatic plant

SOURCES: References for the species mentioned are as follows: northern pike, white bass, and grass carp (Moyle, 2002); silver carp (Schofield et al., 2005); zebra mussel (Cohen and Weinstein, 1998); spiny waterflea and Atlantic comb-jelly (Lowe, Browne, and Boudjelas, 2005); Asian fish tapeworm (Salgado-Maldonado and Pineda-López, 2003); oxygen weed and water chestnut (L. Anderson, UC Davis, personal communication, 2006).

NOTES: Threat ratings are as follows: 1—High likelihood of causing considerable ecosystem/economic damage. 2—Moderate likelihood of causing ecosystem/economic damage. Invasion likelihood in next 25 years: 1—High. 2—Moderate. 3—Small.

to reach 45 million (Johnson, 2005). The most recent update of the California Water Plan assumes that the population may then double—reaching 90 million residents—by 2100 (Department of Water Resources, 2005c). Following trends of the past two decades, much of this growth is expected to occur in the state's inland areas, including the regions bordering the Delta. Such growth will significantly increase both the demand for Delta services and the effects of human activity on the Delta. A growing and seemingly inevitable consequence has been the conversion of Delta farmlands to subdivisions. Estimates prepared by the California State Reclamation Board suggest that as many as 130,000 new homes are currently in the planning stages within the Delta.

Although urbanization can be controlled through regional land use planning mechanisms, there has been little political will to address the issue. Without a dramatic change in state policy, urbanization will powerfully influence the quality of services provided by the Delta. The effects will be seen in two principal ways. First, unlike most other activities in the Delta, urbanization is generally irreversible, barring a catastrophic event like Hurricane Katrina. Once a Delta island is converted to homes, that land use is fixed in place indefinitely; it also promotes the expansion of such services and infrastructure as transportation, utilities, and water systems. Changes in sea level and runoff conditions and the effects of seismicity are unlikely to reverse urbanization. Instead, it is highly likely that after problems caused by these forces, levees will be repaired and raised, and homes will be rebuilt.

Second, urbanization is self-accelerating. Urbanization in one location significantly increases the value of adjacent lands. This, coupled with declining profit margins for farming, will increase the pressure to convert farmlands to subdivisions. This process is already under way in the Delta's "secondary zone"—the upland areas and exempted lowland areas that were slated for development under the 1992 Delta Protection Act (Figure 3.5). In the future, there will be great pressure to build homes within the Delta's "primary zone," despite the act's intent to maintain this low-lying area for agricultural and recreational uses. The increase in number of homes along the perimeter and within the Delta will inevitably shift priorities for Delta management toward flood control and infrastructure to support urbanization. Without major changes in regional land use policy, this shift

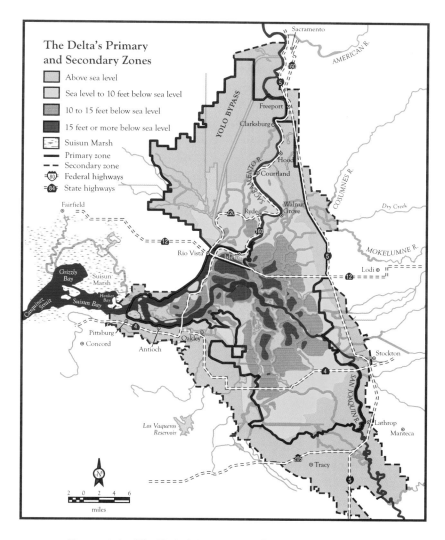

The Delta's Primary and Secondary Zones

- Above sea level
- Sea level to 10 feet below sea level
- 10 to 15 feet below sea level
- 15 feet or more below sea level
- Suisun Marsh
- Primary zone
- Secondary zone
- Federal highways
- State highways

Figure 3.5—The Delta's Primary and Secondary Zones

will come at the expense of habitat protection and other services—such as water quality and water supply—that are important for other parts of California.

Conclusions

The current Delta was developed primarily by creating leveed islands to promote farming in the early days of commercial agriculture. These levees were often constructed with local peat soils and little engineering expertise to protect noncritical land uses—farms that could be restored following any levee failures. Agriculture continues as a major use of the land and as a standard for levee maintenance. However, the use of the Delta both as a conduit for water exports since the 1940s (as described in Chapter 2) and, more recently, as an area of urbanization has increased focus on levee reliability to protect both water quality and urban lands. As described in the next chapter, the Delta's highly altered levee-centric system has been at odds with the aquatic ecosystem, which has experienced a long-term decline in native species and an increased prevalence of undesirable alien species.

The long-term prospects for retaining a levee-centric system for protecting Delta land and water are poor. The existing levee system, even with recently proposed improvements, will be subject to greater probabilities of failure, with sudden and catastrophic consequences for all users of the Delta (Jack R. Benjamin and Associates, 2005). Sea level rise, increasing flood variability, past and continuing land subsidence, earthquakes, and urbanization all contribute to the increasing likelihood of major and multiple levee failures.

When we combine this analysis of the drivers of change in the Delta with a review of our current ecological understanding of the Delta's ecosystem, as described in the next chapter, the current levee-centric strategy for managing the Delta appears unsustainable. Moreover, should the Delta levees fail, the consequences are likely to be sudden and catastrophic for local residents, landowners, Delta species, and water exporters. Currently, the Delta is unsustainable for almost all stakeholders. Responding to the long-term problems of the Delta only after a major catastrophe is unlikely to produce wisely considered or economically prudent policy.

4. The Future of the Delta as an Aquatic Ecosystem

"All truth passes through three stages. First, it is ridiculed. Second, it is violently opposed. Third, it is accepted as being self-evident."

Arthur Schopenhauer

As we saw in Chapter 2, environmental and ecosystem concerns have come to dominate Delta policy, management, and operations in recent decades. This change has come from increased social and political attention to the environment since the 1970s, and it has taken stark legal reality with the listing of several native species as threatened or endangered under the state and federal Endangered Species Acts (Table 2.2). Other federal and state water quality laws (such as the federal Clean Water Act) also influence management of the Delta and estuary. Many aspects of Delta water and land management, from export operations to levee maintenance, are significantly affected by these legal and political concerns. However, these issues are not the only reason for examining the Delta's ecosystem; significant biological issues are also of concern. Invasive species have come to pose expensive challenges to many of the services provided by the Delta. Problems include the collapse of levees from burrowing animals, the clogging of water diversions with alien aquatic weeds, and concerns about the cost and health implications of the physical and chemical means used to control alien species. In addition, recent sharp declines in native species, particularly the delta smelt, indicate the need for attention to biological issues. At the same time, our understanding of the Delta's ecosystem and many of its key species has improved considerably over the last 10 to 20 years, allowing for a more complete analysis of ecosystem problems. This chapter provides an overview of our thinking about the Delta in environmental and ecological terms.

From an aquatic ecosystem perspective, a fundamental conflict exists between two Deltas, namely, the strongly tidal estuarine Delta, which supports a complex ecosystem with a diverse biota, and the agricultural Delta, made up of islands (many subsided) surrounded by high levees. The estuarine Delta naturally fluctuates, both within and across years, between brackish and fresh water. The agricultural Delta created by humans is largely managed as a freshwater system, which provides water for farming and urban areas. Any time that the Delta moves from being a predictable freshwater system toward being a more saline system, major efforts are made to shift it back, by repairing levees, releasing water from reservoirs, reducing water exports, and other actions. As discussed in Chapter 3, it is increasingly evident that a Delta that fluctuates between these states will ultimately win this conflict, as a result of the combined effects of sea level rise, land subsidence, climate change, and levee failures.

The question for this chapter is, "What is likely to happen to the Delta ecosystem as it shifts toward being a more estuarine system in which salinities fluctuate with tides, season, and climate?" Subsidiary questions are: (1) "What habitats need to be abundant in the Delta to favor desirable organisms?" and (2) "What can we do to direct this shift to create an ecosystem that supports desirable organisms?" It is now possible to provide reasonable answers to these questions because of our improved understanding of the ecology of the Delta and the San Francisco Estuary.

Improved Understanding of the Delta Ecosystem

Several basic assumptions on how the estuary operates have proven to be incorrect or only partially correct. Our current understanding of the estuary is based on a series of recent "paradigm shifts" (summarized in Table 4.1 and Appendix A) that should lead to more workable solutions to problems in the Delta. At the same time, it must be recognized that the estuary will continue to change in ways that are difficult to predict, especially as the result of climate change and invasions of alien species. For example, if water temperatures become too warm during the narrow windows of time when delta smelt (*Hypomesus transpacificus*) spawn, their ability to reproduce may be reduced or eliminated (Bennett, 2005).

The present ecosystem is clearly not working well to support desirable organisms, as indicated by the continuing decline of delta smelt, striped bass, and other fish. Because the Delta is always going to have an ecosystem dominated by the combined results of human actions, invasive species, the amount and timing of freshwater inflow, land subsidence, and infusions of toxic materials, the easiest way to assess the nature of desired ecosystem states in the future is to examine how various manipulations will favor key desirable and undesirable species (Table 4.2). Essentially, identifying the species we want in an ecosystem can drive the creation of the most desirable future states of that ecosystem. Throughout this chapter, we focus mainly on the aquatic system but provide some discussion of the terrestrial systems, recognizing that any configuration of the Delta in the future will have to include habitat for key terrestrial species as well, especially overwintering migratory birds (such as waterfowl), neotropical migrants (such as various warblers and thrushes), and sandhill cranes (Table 4.3).

Which Habitats Favor Desirable Organisms?

Views on which organisms are perceived as desirable have changed through the years, but today they include largely (1) native species, especially endemic species (i.e., those native only within a particular area), (2) species harvested for food and sport, including alien species, and (3) species that support the organisms in the first two categories, usually as food, such as copepods and mysid shrimp (Table 4.2). To maintain the Delta as a region that supports these desirable species, especially native aquatic species, there must be habitats with: (1) abundant zooplankton and mysid shrimp, (2) less intrusion of invasive clams, (3) low densities of freshwater aquatic plants, and (4) physical habitat that is diverse in structure and function. To provide these conditions, six basic habitats in the Delta need to be enhanced or maintained: (1) productive, brackish, open-water habitat, (2) brackish tidal marsh, (3) seasonal floodplain, (4) freshwater wetlands, (5) upland terrestrial habitat, and (6) open river channels. These habitats once dominated the San Francisco Estuary. Remnants of these habitats remain and their characteristics can guide restoration efforts, albeit cautiously (Lucas et al., 2002). Overall, a Delta that presents a mosaic of habitats is likely to be the most hospitable to desirable organisms and the most likely to resist invasions by additional alien species. A key to developing such a mosaic is that it would not be stable in either space or time; conditions in each area would change with season and year. Descriptions of the six basic Delta habitats are provided below. Figure 4.1 shows the current locations of these habitats.

Table 4.1

New Understanding of the Delta Ecosystem

New Paradigm	Old Paradigm
1. Uniqueness of the San Francisco Estuary	
The San Francisco Estuary has complex tidal hydrodynamics and hydrology. Daily tidal mixing has more influence on the ecology of the estuary than riverine outflows, especially in the western and central Delta. Conditions that benefit striped bass (an East Coast species) do not necessarily benefit native organisms.	The San Francisco Estuary works on the predictable model of East Coast estuaries with gradients of temperature and salinity controlled by outflow. Freshwater outflow is the most important hydrodynamic force. If the estuary is managed for striped bass, all other organisms, and especially other fish, will benefit.
2. Invasive Species	
Alien species are a major and growing problem that significantly inhibits our ability to manage in support of desirable species.	Alien (nonnative) species are a minor problem or provide more benefits than problems.
3. Interdependence	
Changes in management of one part of the system affect other parts. All are part of the estuary and can change states in response to outflow and climatic conditions. Floodplains are of major ecological importance and affect estuarine function. Suisun Marsh is an integral part of the estuary ecosystem and its future is closely tied to that of the Delta.	The major parts of San Francisco Estuary can be managed independently of one another. The Delta is a freshwater system, Suisun Bay and Marsh are a brackish water system, and San Francisco Bay is a marine system. Floodplains such as the Yolo Bypass have little ecological importance. Suisun Marsh is independent of the rest of the estuary.
4. Stability	
The Delta will undergo dramatic changes in the next 50 years as its levees fail because of natural and human-caused forces such as sea level rise, flooding, climate, and subsidence. A Delta ecosystem will still exist, with some changes benefiting native species. Agriculture is unsustainable in some parts of the Delta.	The Delta is a stable geographic entity in its present configuration. Levees can maintain the Delta as it is. Any change in the Delta will destroy its ecosystem. Agriculture is the best use for most Delta lands.

Table 4.1 (continued)

New Paradigm	Old Paradigm
5. Effects of Human Activities	
Pumping in the Delta is an important source of fish mortality but only one of several causes of fish declines. Entrainment of fish at the power plants is potentially a major source of mortality. Changes in ocean conditions (El Niño events, Pacific Decadal Oscillation, ocean fishing, etc.) have major effects on the Delta. Hatcheries harm wild salmon and steelhead. Chronic toxicants continue to be a problem, and episodic toxic events from urban and agricultural applications are also a major problem.	Pumping in the southern Delta is the biggest cause of fish declines in the estuary. Fish entrainment at power plants is a minor problem. Changes in ocean conditions have no effect on the Delta. Hatcheries have a positive or no effect on wild populations of salmon and steelhead. Chronic toxicants (e.g., heavy metals, persistent pesticides) are the major problems with toxic compounds in the estuary.

Table 4.2

Important Aquatic Species and Habitat Conditions That Improve Their Abundance

Species	Desirability	Description	Salinity[a]	Temperature (°C)	Flow	Rearing Habitat
Delta smelt	+++	Threatened species, endemic	Fluctuating	< 20°	Tidal	Open water, pelagic, brackish
Longfin smelt	+	Declining species	Fresh-marine	< 16°	Tidal	Open water, pelagic, marine
Splittail	+	Endemic	Brackish-fresh	< 24°	Tidal	Brackish tidal marsh
Tule perch	+	Native, declining	Fresh-brackish	< 22°	Tidal, river	Tidal marsh, river edge
Striped bass	++	Sport fish, declining	Fluctuating	< 25°	Tidal	Open water, pelagic, brackish
White sturgeon	+	Sport fish, declining	Brackish-marine	< 20°	Any	Bottom, open bay
Green sturgeon	++	Threatened species	Fresh-marine	< 20°	High river for spawning	River, then marine
Chinook salmon[b]	+++	Endangered to commercially fished	Fresh-marine	< 20°	Tidal, river currents	Shallow edge and flooded marsh
Large estuarine copepods	+++	Important in food webs	Fluctuating	Depends on species	Tidal	Open water, pelagic
Mysid shrimp	+	Important in food webs	Fluctuating	< 20° ?	Tidal	Open water, pelagic
Diatoms	++	Basis for food webs	Various	Various	Tidal	Open water
Largemouth bass	−	Alien predator, game fish, indicator	Fresh	< 30°	None, low	Backwaters, sloughs
Asiatic clam	−	Alien filter-feeder	Fresh	< 35°	River, tidal	River channels, flooded islands
Overbite clam	− − −	Alien filter feeder	Brackish	< 23°	Tidal	Suisun, San Francisco Bays

Table 4.2 (continued)

Species	Desirability	Description	Salinity	Temperature (°C)	Flow	Rearing Habitat
Brazilian waterweed	– – –	Alien plant pest	Fresh	< 35°	None-low	Delta sloughs
Water hyacinth	–	Alien plant pest	Fresh	< 35°	None	Delta sloughs

NOTES: +/– indicates desirability of species to humans as seen by how likely the species are to influence management decisions (+ positive, – negative, with the strength of the desirability indicated by the number of + and – signs). All + species are declining, and all – species are abundant or increasing. Temperature, salinity, and flow represent preferred conditions.

[a] Fluctuating salinities means that the salinities will change enough on an annual and interannual basis to discourage undesirable nonnative species.

[b] There are four runs of salmon (winter, late fall, fall, and spring) with different status and habitat requirements.

Table 4.3

Selected Important Terrestrial/Upland Species for Which Changes in the Delta, Suisun Marsh, and Surrounding Areas Will Cause Changes in Abundance

Species	Desirability	Importance	Delta	Suisun	Upland/ Agricultural Areas	Riparian Areas	Notes
Wintering waterfowl	+++	S, R, D, B	xxx	xxx	xx	xx	50+ species
Neotropical migrant birds	++	R, D, B	xx	xx	xxx	xxx	Many species, including three species listed under CESA
Swainson's hawk	++	E, S, R	xx		xxx	xx	
Sandhill crane	+++	E, S, R	xxx		xxx	x	Major wintering population
California clapper rail	++	E, R		xx			Requires tidal marshes
Black rail	+	E, R	x	x			
Yellowbilled magpie	+	S, D, R	xx	x	xxx	xx	Decline from West Nile virus
Salt marsh harvest mouse	++	E		xxx			
Beaver	–	N, R	xx	x		xx	Burrows into levees
Muskrat	– –	N, I	xx	x	x	xx	Burrows into levees
River otter	+	S, R	xx	xxx		xx	Major population
Mexican freetail bat	+	S, R, B	xx	x	xx	xx	Large population, eats pest insects
Giant garter snake	+	E	xx		x	xx	Listed
Fairy shrimp	+	E	x		xxx		Four listed species under ESA and CESA; in vernal pools (special habitat)
Valley elderberry longhorn beetle	++	E	xx	x	x	xxx	*De facto* protection for elderberry bushes; may be delisted

Table 4.3 (continued)

Species	Desirability	Importance	Delta	Suisun	Upland/ Agricultural Areas	Riparian Areas	Notes
Mosquitoes	– –	N	xxx	xxx	xxx	xxx	Several species; spread West Nile and other diseases
Tules	++	S, B	xxx	xxx	x	xx	*Scirpus* species; habitat, bank protection, etc.
Fremont cottonwood	+	S, B	xx		x	xxx	Major riparian tree, important for birds, etc.
Local endemic plants	+	E, D	xx	xxx	xxx	xx	Highly localized with special requirements
Perennial pepperweed	– –	N, I	xx	xx	xx	xx	Representative of invasive alien plants

NOTES: +/– indicates desirability of species to humans as seen by how likely the species are to influence management decisions (+ positive, – negative, with the strength of the desirability indicated by the number of + and – signs). Importance: E (listed as threatened/ endangered by state or federal agencies), D (declining), S (symbolic, charismatic, and emblematic of region), R (recreation, hunting, birdwatching, etc.), N (nuisance species), I (introduced species), B (ecologically beneficial species). The number of xx's in major habitat area indicates importance of the habitat to the organism.

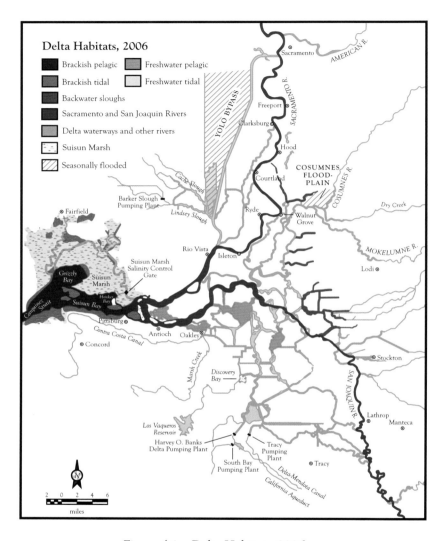

Figure 4.1—Delta Habitats, 2006

Productive Brackish Open-Water Habitat

For the past 20 to 25 years, the greatest concern over declining numbers of fish that depend on the Delta has been for open-water (pelagic), plankton-feeding fish, mainly delta smelt, longfin smelt, and striped bass. Their long-term decline has apparently accelerated since 2001, increasing concern for the viability of their populations and those of other pelagic fish.[1] This decline is tied in part (but by no means entirely) to the shift in the food web of Suisun Bay and the Delta. Previously, most energy and carbon flowed through pelagic zooplankton and fish; currently, most energy and carbon instead flow through the alien overbite clam (*Corbula amurensis*), which became established in the region in 1986 (Carlton et al., 1990).

Historically, Suisun Bay was the principal brackish water region where most open-water habitat existed. It was without abundant clams (except in dry years when marine clams invaded) and therefore supported abundant diatoms (a type of algae or phytoplankton), which were fed on by zooplankton (mainly *Eurytemora affinis*, a copepod), which in turn were fed on by both small plankton-feeding fish (e.g., delta smelt) and mysid shrimp (mainly *Neomysis mercedis*). The mysid shrimp then became a major item in the diets of larger planktivores, especially longfin smelt and juvenile striped bass. But with the invasion of the brackish water tolerant overbite clam, these food organisms became greatly depleted, presumably reducing the growth and survival of the planktivores. Thus, open-water habitat still exists, but its productivity is funneled more into clams than into desirable fish.

As productive open-water habitat has diminished in brackish water areas, other areas favorable to pelagic organisms have been reduced as well. This loss is mainly the result of the Brazilian waterweed (*Egeria densa*) and other submerged aquatic vegetation, which have invaded freshwater sloughs, channels, and flooded islands of the Delta (Brown, 2003). Waterweed grows in dense mats in shallow water (< 3 m) along the channel edges and can completely choke shallow quiet water habitats during the warmer months. These plants slow the flow of water and retain sediments, nutrients, and other materials from the water column; consequently, the water tends to be clearer. These more transparent waters support populations of alien invertebrates and fish, including centrarchids, mainly largemouth bass, bluegill, and redear sunfish. In contrast, the more open, less transparent habitats in the Delta are more likely to support populations

---

[1] http://science.calwater.ca.gov/pdf/worksho ps/POD/IEP_POD_Panel_Review_Final_010606_v2.pdf. For a graph showing trends in abundance indices of key pelagic species, see Figure 1.3.

of striped bass, delta smelt, Chinook salmon, and splittail (Nobriga et al., 2005).

Generally, where Brazilian waterweed is abundant, open-water habitat is reduced and alien fish and invertebrates dominate, conditions mostly undesirable from an ecosystem perspective (Brown, 2003; Nobriga et al., 2005). The bass (and other warm-water fish) support fisheries, but these fisheries do not depend on the estuary for their existence (as do fisheries for striped bass, salmon, and splittail). Where currents are too strong for Brazilian waterweed to become established, freshwater channels may support dense populations of the Asiatic clam (*Corbicula fluminea*) which can strip the water column of plankton, reducing food supplies for pelagic fish. This is especially true today in the southern Delta, where the Asiatic clam is abundant in the San Joaquin River channel.

These changes mean that estuarine-dependent pelagic organisms, such as striped bass, have seen a loss of habitat in both freshwater and brackish water. The key to restoring the desirable pelagic species is to recreate habitats that have a high variability in nonbiological (or "abiotic") factors such as salinity, channel flows, depth, and water clarity (Nobriga et al., 2005; Lopez et al., 2006). This is the kind of estuarine habitat that once dominated many Delta channels and Suisun Bay: open-water areas that varied sufficiently in salinity from fresh to moderately salty (roughly 8–10 parts per thousand (ppt)) seasonally or across years and often had strong tidal currents and low water clarity.[2]

In areas where such conditions return, it is unlikely that the overbite clam, Brazilian waterweed, or the Asiatic clam will be able to persist. It appears that moderate salinities during the summer growing season will exclude Brazilian waterweed. The Asiatic clam may require salinities exceeding 13 ppt for complete exclusion but the species is rarely abundant where salinities exceed 5–6 ppt for extended periods of time (Morton and Tong, 1985). Unfortunately, the biggest problem species in brackish water, the overbite clam, can live and reproduce in water ranging from fresh to 28 ppt, at temperatures of 6°C to 23°C (Parchaso and Thompson, 2002). Like many clams, its growth and reproduction are limited by food supply,

---

[2]As a rough guide, seawater is 35 ppt and fresh water is less than 3 ppt. Drinking water is less than 1 ppt.

but this clam is large enough and lives long enough (two to three years) so that it can survive many weeks with limited food (Parchaso and Thompson, 2002).[3] Nevertheless, the overbite clam is highly stressed when exposed to fresh water (Werner, 2004) and has not colonized areas in the estuary that are fresh for extended periods of time, despite being physically able to do so. This suggests that annual exposure to fresh water for three to six months may limit its ability to invade some areas.

Today, the best example of habitat with low numbers of these alien species is Suisun Marsh, especially in Nurse Slough (R. E. Schroeter, UC Davis, personal communication, 2006). This turbid habitat, with few clams, contains abundant phytoplankton and zooplankton and thus is favorable for rearing small estuarine fishes such as delta smelt and juvenile striped bass. Essentially, this habitat has enough variability in abiotic conditions, especially salinity, that undesirable populations of both freshwater and brackish water organisms are inhibited.[4] The most likely location of restored habitat of this nature would be on flooded islands close to sources of both salt water and fresh water (e.g., Sherman Island, Twitchell Island). Alternatively, undesirable alien species could be excluded by keeping islands completely enclosed by levees but adding gates that would allow free access to tidal flows in most years. If gated, these pelagic habitat islands could be drained and dried as a control measure for invasive species when necessary (Table 4.4).

---

[3]Overbite clams can persist in fresh water because they can burrow into sediments, which can retain salts for long periods of time, and then clamp their valves together until good conditions return. "So a Corbula living in the sand can simply burrow down, crack its valves for a little freshening periodically and live as long as the water doesn't drop below its oxygen limit or until it runs out of energy stores" (J. Thompson, U.S. Geological Survey (USGS), personal communication, May 2006). Nevertheless, most overbite clams residing in lower Suisun Slough were killed during the winter of 2005–2006, presumably because of continuous freshwater flows from Cordelia Slough, which receives water from nearby creeks. Clams survived, however, in the reach of Suisun Slough immediately above the mouth of Cordelia Slough, which lacked the heavy freshwater influx (R. E. Schroeter, UC Davis, personal communication, 2006).

[4]What may be as important as variability per se is the suddenness of change; conditions, especially salinity, that change abruptly (over a few days) may eliminate undesirable organisms more effectively than more gradual change.

Table 4.4

Likely Responses of Populations of Common Delta Fish and Shrimp to Increases in Three Salinity Regimes in a Large Open-Water Environment

Species	Fresh	Brackish	Fluctuating
Delta smelt	– –	–	+
Longfin smelt	–	–	+
Striped bass[a]	–	–	++
Splittail	0	+	++
Tule perch	+/–	?	+
Prickly sculpin	–	0	+
Hitch	+?	0	0
Blackfish	+	0	0
Fall-run Chinook	+/–	+/–	+/–
Spring-run Chinook	+	+	+
Winter-run Chinook	+	+	+
Steelhead	0	0	0
White sturgeon	0	+	0
Largemouth bass[a]	++	0	–
Lepomis spp[a]	++	0	–
Inland silverside[a]	++	+	+
American shad[a]	0	0	0
Threadfin shad[a]	+	0	+
Shimofuri goby[a]	0	+	+
Yellowfin goby[a]	0	+	+
Golden shiner[a]	++	–	–
Mosquitofish[a]	++	+	0
Siberian prawn[a]	–	+	++
Mysid shrimp	0	+	+

NOTES: For definitions of symbols, see Table 4.2. Salinity in this case is the indicator of the changed environment; changes in water clarity, temperature, and depth would also influence fish populations. A freshwater habitat would essentially resemble present-day Franks Tract and Mildred Island. A brackish water habitat would be like present-day Suisun Bay. A fluctuating salinity environment would be most like portions of Suisun Marsh.

[a]Indicates non-native species. 0 = no change.

Brackish Tidal Marsh

Brackish tidal marsh is the main habitat along the sloughs of Suisun Marsh, in the unleveed portions of Suisun Marsh, and in marshes along the edge of Suisin Bay. This ecosystem was once much more extensive in Suisin

Marsh, Suisun Bay, and the lower Delta. Brackish tidal marsh is typically shallow (< 2 m at high tide), cool (< 20°C), turbid (transparency < 35 cm), and complex in structure, with a strong tidal influence (Matern, Moyle, and Pierce, 2002; Brown, 2003). Such habitat is important for rearing desirable fish, especially splittail, juvenile striped bass, and perhaps juvenile Chinook salmon. Not only are fish in general more abundant in the unleveed sloughs, but the proportion of native fish also tends to be high (R. E. Schroeter, personal communication, 2006). Such areas also are presumed to be an important source of nutrients for adjacent channels and bays. Areas inundated by tidal water for only short periods support vegetation important for such threatened species as salt marsh harvest mouse, black rail, and clapper rail.

With sea level rise, this habitat will expand in Suisun Marsh, as levees eventually overtop and breach. The depth of the habitat will depend on how much subsidence occurs before the inevitable flooding takes place and on how much the growth of submerged vegetation keeps up with sea level rise. Ideally, some shallow channels in the marsh will continue to have characteristics that exclude the overbite clam and favor native fish, through the input of fresh water from the Sacramento River, local runoff, and, perhaps, tertiary treated sewage from the Suisun-Fairfield urban area. If we recognize the inevitability of sea level rise, it should be possible to maximize its benefits or control its effects, by planning for a "new" brackish Suisun Marsh.

Seasonal Floodplain

Recent studies show that seasonally flooded habitat in and just above the Delta (i.e., Yolo Bypass, Cosumnes Preserve) is important for spawning splittail and for rearing juvenile salmon and other fish (Sommer et al., 2001a; Crain, Whitener, and Moyle, 2004; Moyle et al., 2004; Moyle, Crain, and Whitener, in press). The Yolo Bypass is unique as a "flow through" system, in which water has a limited "residence time" (i.e., it moves through the bypass relatively quickly). As a result, it floods on an irregular basis (when water spills over the Fremont Weir) and drains quickly. Much of the invertebrate biomass is chironomid midges, which can persist (as eggs) in dry soil.

The most productive floodplain habitat for fish outside the Yolo Bypass is covered with annual vegetation and is flooded with river water from roughly early February through April. In contrast to the Yolo Bypass, the water in these areas often drains slowly, so has a high residence time, allowing it to develop dense populations of zooplankton. The best places to create and maintain such habitat (e.g., expanded Cosumnes Preserve, Cache Slough region, lower San Joaquin River) need to be actively managed to maintain a habitat mosaic and to make sure that flooding occurs on at least part of the available habitat each year. These areas can also be important foraging and roosting areas for migratory waterfowl.

Freshwater Wetlands

Much of Suisun Marsh and parts of the Delta (e.g., Cache Slough region) are managed directly or by default as freshwater marshes. Such marshes are important for an array of plants and animals, especially waterfowl and shorebirds. There are several types of these wetlands, with distinctive characteristics, that presumably all need to be maintained. As the area of freshwater wetland shrinks in Suisun Marsh, more freshwater wetlands may have to be created on Delta islands currently devoted to agriculture, especially if waterfowl habitat (and hunting) is to be supported at present levels. These islands could follow the models proposed by Delta Wetlands Corporation, which have wide levees that slope toward the interior, supporting riparian vegetation and interior water levels that are managed for waterfowl (or water storage).[5]

Upland Terrestrial Habitat

Agricultural areas, especially those islands on which corn and rice are grown, can be important foraging areas in winter for sandhill cranes, migratory waterfowl, and raptors such as Swainson's hawk. Presumably such areas will continue to exist in parts of the Delta that lie at or above sea level. However, this habitat is prone to urban development. To maintain adequate

---

[5]The Delta Wetlands project is a proposal to use two islands in the central Delta (Bacon and Webb) as freshwater storage facilities and two others as waterfowl habitat. It is one of five surface storage projects identified in the CALFED Programmatic Record of Decision (CALFED, 2000a).

areas of this habitat, substantial tracts (e.g., Staten Island) will have to be managed, often behind levees, with wildlife as the highest priority.

Open River Channels

Delta channels, especially those leading to flowing rivers, must be maintained as migratory corridors for salmon, steelhead, lamprey, splittail, delta smelt, and other fish. Ideally, fish migration corridors should also minimize the risk of entrainment in the pumps in the southern Delta. These channels also need to provide juvenile rearing habitat along their edges and offer connectivity between spawning and rearing areas (e.g., for splittail, between floodplain spawning habitat and brackish tidal marsh rearing habitat). The present configuration of the Delta, especially the southern Delta, results in complex flow patterns through the channels that presumably confuse migratory fish going both upstream and downstream. Channel configurations need to be reconstructed in ways that resemble historical conditions—that is, with more natural spatial patterns with fewer straight lines and more dendritic, or branchlike, patterns (J. Burau, USGS, personal communication). These channels also need to be managed in ways that discourage alien species.

How Can We Create a Delta That Supports Desirable Organisms?

The crisis brought on by the continuing pelagic organism decline, especially delta smelt, has led to the realization that the Delta ecosystem is not providing for the needs of key organisms. The growing recognition that major changes to the Delta will occur as the result of the factors discussed in Chapter 3 is also forcing a reexamination of the future of the Delta ecosystem. In addition, we now know that many of our basic assumptions about how the Delta operated as an ecosystem that were used in planning in the past were wrong or misguided (Table 4.1 and Appendix A). Taken together, these realizations provide both the motivation and the opportunity to rethink how we might manage the Delta's ecosystem, using guidelines that follow.

Given the inevitable changes that will occur to the Delta ecosystem, our choice is either to respond to each change as a disaster or to plan for it as an opportunity to create more predictable and productive environments for fish and wildlife. Some key features of a carefully planned effort at controlling change to favor desired organisms include (1) tying the Delta to adjacent ecologically important areas, (2) creating island and channel habitat diversity by reengineering Delta planforms to enhance dendritic channel patterns that support various habitats (particularly in terms of salinity and water residence time), (3) preventing the "hardening" of secondary Delta lands by urban development, and (4) improving connectivity between rivers and parts of the Delta.

Tie the Delta to Adjacent Areas

Much of the discussion of the Delta ecosystem focuses on the central and southern Delta because these areas have significant subsidence problems and major, immediate connections to the SWP and CVP pumps. From an ecological point of view, it is unclear what can or will actually be done to islands in these areas to benefit the species of concern, given the high likelihood of uncontrolled flooding (discussed in Chapter 3). We need therefore to look to areas adjacent to the Delta to provide most of the desired ecological functions. It is also quite likely that money invested in these adjacent areas will produce a bigger return in ecological value on a per dollar basis than money spent on interior Delta projects. Some key areas include:

1. **Cache Slough region.** This area, to the north, adjacent to the Yolo Bypass, is within the legal boundaries of the Delta but is rarely discussed in a Delta context, in part because what happens there has little effect on the delivery of fresh water via the pumps of the southern Delta. Yet it has large tidal excursions (much of the tidal water moving up the Sacramento River channel winds up there), a complex, branching channel pattern, and is a known spawning and rearing area for delta smelt and probably for other native fish as well. It is the outlet for water draining from the Yolo Bypass, with potential major interactions ranging from exporting nutrients to rearing juvenile salmon (Sommer et al., 2001a and 2001b). Arguably, this region is most like the historical Delta, although many of its channels have been leveed or otherwise altered. A "natural" levee failure experiment exists there now (Liberty Island, which flooded in 1998) and much of the

land is in private ownership. It also has the intake for the North Bay Aqueduct (in Barker Slough), which may constrain some uses.

2. **Yolo Bypass.** The Delta doubles in size when the Yolo Bypass is flooded. The problem is that the bypass floods only erratically and not always at times optimal for fish and birds. The bypass presents some major opportunities for ecosystem manipulation (e.g., by gating the Fremont Weir), which are currently under discussion (Department of Fish and Game, 2006). It is also a major spawning and rearing area for splittail and other native fish, a rearing area for juvenile salmon, and a potential source of nutrients for Delta food webs (Sommer et al., 2001a and 2001b). This region could act as a major interface with the Delta ecosystem, especially in the Cache Slough region, a role that will likely grow in importance, both through deliberate manipulations and through the increased frequency of flooding as a result of climate change.

3. **Van Sickle Island/Southern Suisun Marsh.** Van Sickle Island is a major marshy island that borders the west side of upper Montezuma Slough (by the tidal gates) and the south side of Suisun Bay, where the Sacramento River enters. Its levees failed in several places during the winters of 1997–1998 and 2005–2006, but they were fixed by DWR to protect infrastructure around the Roaring River that helps to keep salt water at bay.[6] This infrastructure is the water delivery system that maintains the interior marshes as freshwater systems for duck hunting clubs. One potential negative effect of allowing Van Sickle Island to flood is that this may increase the likelihood of highly saline water arriving at the pumps of the southern Delta. Nevertheless, Van Sickle Island has high potential as a place to create a large expanse of brackish tidal marsh, a desirable feature that may be inevitable as sea level rises. The potential negative effect on water delivery might be lessened if the island were breached on the Montezuma Slough side, with south-side levees being maintained, before the system was inundated naturally.

---

[6]DWR took this step even though these are private levees, not "project" levees under state and federal responsibility.

4. **Cosumnes/Mokelumne River confluence area.** The Cosumnes River preserve is a floodplain demonstration area, relatively small, but important for fish spawning and rearing (Moyle, Crain, and Whitener, in press). There are opportunities both within the preserve area and nearby for expanding the floodable lands and creating more upland habitat useful for sandhill cranes, waterfowl, and other species of interest.

5. **Upland agricultural areas.** Sandhill cranes and waterfowl need these farmland areas, preferably planted in corn, for winter foraging. Much of this habitat is on islands that could or will flood (e.g., Staten Island). However, upland areas around the Delta are increasingly turning into housing tracts and vineyards. This trend needs to end if habitat for cranes and waterfowl is to be maintained. This is especially important as heavily subsided islands become submerged or converted to other uses.

Create Island and Channel Habitat Diversity

If we want habitat heterogeneity, then we should consciously choose the types of island and channel habitats we want and figure out how to achieve the right balance among them. This process would involve managing island levees and land uses, as well as reengineering some Delta channels to create a more naturally diverse dendritic channel structure, which would allow for greater variability in salinity, residence time, and flow velocities across the Delta (J. Burau, personal communication, 2006). Of course, the possibilities for restructuring the system will depend on the nature of the cross-Delta water delivery system. Here are some possible alternatives for island and channel management:

1. **Natural pelagic habitat.** This would consist of islands or sections of islands in the western Delta (i.e., Sherman, Twitchell, Bradford, Jersey) in which strategic levee breaches could cause strong tidal excursions, allowing salinity fluctuations that inhibit overbite clam, Asiatic clam, Brazilian waterweed, and other undesirable species. Basic island configuration could be maintained by specially designed levees, if desired, but it might be possible to just let one or two islands revert to open water without levees. Without significant effort, however, many

subsided islands will become warm-water fish habitat like Franks Tract or Mildred Island, described below.

2. **Controlled pelagic habitat.** These areas would be modeled on the proposed Delta wetlands project and would feature sloping interior levees supporting riparian forest and tule beds.[7] They would have gates in several places to regulate inflow and outflow. An ideal feature would be the ability to dry them completely when undesirable invasive species become too abundant. If strategically placed, islands with sufficient area and depth might be used to regulate salinity or outflow in extreme situations (e.g., levee failures on other islands). One advantage of this kind of management is that options for various ecological and water supply uses would be kept open.

3. **Wildlife habitat.** These islands could also be maintained for ducks and other waterfowl, as in the Delta Wetlands model. They would be flooded only enough to produce duck habitat, which includes some wildlife-friendly farming, and would presumably be dry in summer, except for recreational ponds. Waterfowl production and hunting opportunities are likely to decrease in Suisun Marsh, as a result of flooding by salt water from sea level rise and deliberate manipulations. Hunting could shift from Suisun Marsh to some Delta islands, where new hunting clubs could be established. This shift would allow for opportunities to create more tidal habitat in Suisun Marsh. This option assumes, of course, that subsided islands with large, inward-sloping levees would be able to resist flooding from sea level rise and that a source of fresh water would be available for wildlife habitat. Much would depend on the amount and rapidity of sea level rise and on the design and operation of the interior Delta.

4. **Warm-water fish habitat.** Franks Tract and Mildred Island are examples of warm-water fish habitats and originated as subsided islands that have been "let go." They have become heavily invaded by alien species from plants to invertebrates to fish, but they do have such recreational benefits as boating and fishing. The location and size of

---

[7] Here, we suggest an alternative use of flooded islands—for habitat instead of freshwater storage—using the same basic technology of sloped and rocked interior levees.

such open-water areas in the Delta could make a big difference both in Delta tidal circulation and in the timing and frequency of saltwater fluctuations.

5. **Agricultural islands.** Some of the least subsided islands could be maintained indefinitely for wildlife and Delta-friendly agriculture. A key would be to promote agricultural practices that discourage urbanization and prevent—or even reverse—further subsidence. One focus for the development of such islands could be sandhill crane and Swainson's hawk foraging areas.

Prevent Hardening of Adjacent Upland Areas

When upland areas around the Delta become urbanized, are turned into vineyards, or become devoted to other uses that greatly increase land values, land use choices diminish. "Hardened" areas are also likely to have increased human use, and this change may have significant consequences for wildlife. For example, if Staten Island and other Delta islands that are used by sandhill cranes for foraging become submerged, the cranes will need similar agricultural land elsewhere—and hardened areas will be unable to provide it.

This is largely a planning issue, and big development forces are arrayed against the maintenance of low-value farm crops (see Chapters 3 and 5). But the value of these upland areas to wildlife, including endangered species, should be emphasized. Rather than an area of urban development, the Delta could be considered open space and a benefit to citizens of nearby urban areas, from Sacramento to Stockton to San Francisco.

Improve Connectivity

In any proposed changes, the importance of Delta channels for upstream and downstream migrating fish has to be kept in mind. Clear migration routes to the Sacramento and San Joaquin Rivers, as well as to the Mokelumne and Cosumnes Rivers, must be maintained and enhanced. Potentially, a redesigned Delta could improve connectivity in a number of ways: by reducing exposure of fish to entrainment in the pumps in the southern Delta and other agricultural, urban, and power plant diversions; by better management of barriers and gates on Delta channels; by rebuilding key channels to improve passage and water movement; and by providing rearing habitat for juvenile fish. Improving connectivity is clearly not an easy task in the effort to balance water supply and ecological needs in a changing Delta. For example, in the present Delta, the delta smelt and Chinook salmon have different, and at times opposing, needs.

Research Needs and Potential Experiments

Management of the Delta as an ecosystem should be driven by the best scientific information available. Despite considerable new information, a great deal of uncertainty remains about the effects of various management actions. Nevertheless, there is a growing consensus that major change is going to happen, whether we like it or not. Because there is never enough information to make decisions with absolute certainty, a synthesis of existing information is needed to reduce decisional paralysis. Here are some suggestions.

1. **Commission an overview.** Given the great increase in knowledge of the system in the past 15 years, it would be useful to have a new, overarching study of the ecology of the estuary, along the lines of Herbold and Moyle (1989) and Herbold, Jassby, and Moyle (1992), beyond just the open-water system (Kimmerer, 2004).
2. **Examine invasive species.** A recently compiled database on invasive species in the Delta (Light, Grosholz, and Moyle, 2005) begs for analysis of species interactions, potential problem species in response to Delta changes, and predictions of the nature of potential future invaders.
3. **Develop predictive models.** The interactive effects of changing salinity, temperature, depth, water clarity, and flow on key alien species such as Brazilian waterweed, overbite clam, Siberian prawn, and Asiatic clam in particular should be studied.
4. **Pursue synthetic studies.** These studies should focus especially on how to manage the Cache Slough region and Suisun Marsh for desirable species, as sea level rises and climate changes. The Cache Slough region also needs basic ecological studies.
5. **Perform hydraulic modeling.** Analyze whether it is possible to manage selected islands as open-water systems to favor desirable pelagic organisms (delta smelt, striped bass, etc.)—and if so, how.

6. **Develop experimental islands.** A factor that inhibits taking action to convert Delta islands to different uses is uncertainty: What happens in reality when we breach levees or allow an island to be flooded? One way to reduce uncertainty is to develop experimental islands. This is being done today at Dutch Slough on the southwestern edge of the Delta, although funding limitations are reducing options and monitoring (B. Herbold, U.S. EPA, personal communication, 2006). Sherman Island also has potential for experimentation, because of its shallowness and key location near the lower apex of the Delta. It could be segmented into smaller "islands" with different experimental flooding regimes (J. Cain, Natural Heritage Institute, personal communication, 2006).

Some of this research might be accomplished by traditional agency and academic efforts. However, there will be an increasing need to integrate research efforts to make faster improvements in our understanding and to focus additional research efforts more intently on remaining uncertainties. The efforts of the CALFED science program in this area remain embryonic and are not particularly integrated. Greater funding and much greater scientific leadership will be needed if we are to take an aggressively adaptive approach to management.

Conclusions

The Delta ecosystem has been changing rapidly and often unpredictably for the past 150 years, a trend that is likely to accelerate unless we take action to control the change as much as we can. Ultimately, the rate of change may slow down even if we do nothing but respond to emergencies. However, the resulting Delta system is likely to have many undesirable features and species and to be missing many of the species we regard as important today. Such an outcome is not inevitable, though. There are reasonable steps that can be taken to restore Delta habitats to more desirable, variable conditions in terms of flow and water quality, conditions that would better support desirable species and disrupt the establishment of invasive species.

The approach outlined here represents a new and different scientific understanding of how the Delta and its ecosystem function. As will be seen in later chapters, our improved understanding of the Delta's ecosystem leads to the consideration of very different land and water management alternatives and to new conclusions for Delta policy and management. New and more promising alternatives can be designed to take advantage of this improved understanding.

Before exploring these alternatives, we provide some background on recent Delta policymaking (Chapter 5) and then assess the ability of water users and the larger water supply system to adjust to changes in Delta water management policies (Chapter 6). In the end, it is desirable to have solution alternatives that support as many as possible of the Delta's current services.

5. A Crisis of Confidence: Shifting Stakeholder Perspectives on the Delta

"The greatest challenge . . . is stating the problem in a way that will allow a solution."

Bertrand Russell

By December 2004, the decade-old truce between water users and environmental groups, forged at the beginning of the CALFED process, was all but over. This truce—epitomized by the CALFED motto that "everyone would get better together"—had always been a fragile one, with continuing differences over priorities for the Delta within the CALFED investment portfolio. Disagreements had escalated over the course of 2003, as conflicts arose over a water user proposal to increase Delta export levels. Then, through the summer and fall of 2004, concerns surfaced in quick succession over the viability of two central CALFED components: the stability of the levee system and the protection of native fish. Several months after a highly publicized levee failure on Lower Jones Tract drew attention to Delta flood risks, a new analysis of the systemic long-term risks to Delta levees was reported at the October CALFED Science Conference (Leavenworth, 2004a, 2004b, 2004c). Meanwhile, routine fall fish surveys registered sharp declines in several pelagic species, including the threatened delta smelt.[1]

The CALFED 10-year finance plan, released in early December 2004, increased the intensity of this storm. The $8 billion plan drew immediate fire from legislators and stakeholders, who criticized it for being either unrealistic or unfair (Taugher, 2004). The plan proposed to substantially increase financial contributions from the federal government and water users, both of which had been much lower than anticipated when the CALFED ROD was signed in 2000 (CALFED, 2004a). In a sense, the 10-

---

[1] Figure 1.3 shows the trends in abundance of several key pelagic species.

year plan merely articulated the weaknesses in CALFED's finances that had already become apparent: The federal government was a less enthusiastic donor than CALFED architects had hoped; implementing the "beneficiary pays" principle to elicit water user contributions was proving elusive; and state bond funds, which had taken up the slack, were running out.

The storm gathered strength over the course of 2005. Much of its fury was directed at the CALFED governing and implementing bodies. The legislature slashed the program's budget, and the governor's office called for three multifaceted audits to look at finance and governance questions. An interagency POD task force was set up to investigate the reasons for the pelagic organism decline.[2] Meanwhile, in a vote of no confidence in the collaborative processes of the preceding decade, the environmental community filed lawsuits against the federal government on two biological opinions related to Delta exports. About this time, Hurricane Katrina struck in New Orleans, reinforcing concerns over Delta levees and highlighting that levee expenditures under CALFED had been too modest to offer much new protection. In November, DWR began a round of briefings stressing the dire consequences of a catastrophic levee failure for water supply, farmland, homes, and infrastructure (Thompson, 2005b; Snow, 2006).

The audit of CALFED's governance structure revealed weaknesses that had prevented the effective implementation and oversight of its programs (Little Hoover Commission, 2005) and put institutional reform of CALFED on the administration's and legislature's agenda. The financial review confirmed the disproportionate contributions of the state, which covered 41 percent of the $2.5 billion in total expenditures in the first four years, compared to only 10 percent by the federal government (Department of Finance, 2005). Although contributions by water users and local water agencies amounted to a hefty 49 percent, the majority of these funds were local matches for local water supply projects (groundwater banking, conservation, and recycling investments) that would probably have gone forward anyway.

---

[2]The POD team's early reports suggested a complex set of reasons for the collapse of the open-water species (Weiser, 2005).

As Chapters 3 and 4 have shown, future approaches to the Delta will need to revisit CALFED's assumptions about the long-term sustainability of the levee system and its approaches to ecosystem protection. Moving in this direction calls not only for new science but also for new agreements among various stakeholders. In this chapter, we examine current stakeholder perspectives on problems in the Delta, drawing on press accounts, other published documents, and conversations with over 40 stakeholders representing water users, environmental groups, and various in-Delta interests.[3] This review suggests that fashioning agreement on a new vision for the Delta may be even more challenging now than when the CALFED process was launched in the mid-1990s.

Shifting Stakeholder Perspectives

The recognition of new problems in the Delta has reinforced various stakeholders' concerns about the CALFED program's ability to address their primary interests. Each group's interests correspond to one or more of the four broad goals laid out in the CALFED ROD: water quality, ecosystem support and restoration, water supply reliability, and levee stability.[4] Whereas environmental groups and agencies have been principally concerned with the CALFED's ecosystem goals, urban and agricultural water exporters in the Bay Area, the San Joaquin Valley, and Southern California have focused on the program's water supply reliability objectives, with water quality as a secondary concern. By contrast, for water users that draw directly from the Delta—including Delta agriculture and the Contra Costa Water District—managing water quality (particularly salinity) has been a primary objective. Delta farmers had been

---

[3]For a list of persons consulted, see Appendix B. Because some individuals preferred not to be quoted, we use the information gathered from these conversations to inform the discussion here. The reader is referred to press accounts for public statements by various stakeholders.

[4]Specifically, the CALFED ROD stated the water reliability objective as follows: "Reduce the mismatch between Bay-Delta water supplies and current and projected beneficial uses dependent on the Bay-Delta system" (CALFED, 2000a, p. 9).

the only consistent advocates of the CALFED levee program before current, increased recognition of the wider consequences of levee failure.

Levee Problems Draw Attention to a Broad Range of Delta Land Uses

The new spotlight on levees has been of particular concern to interests within the Delta, and it has drawn attention to some stakeholders overlooked in earlier CALFED processes: cities and towns with current or planned development behind Delta levees and various infrastructure providers (e.g., Caltrans, Pacific Gas & Electric (PG&E), East Bay Municipal Utilities District (EBMUD), railroads, ports) whose investments depend on the stability of Delta islands. The increasing urban and recreational value of land in the Delta also has brought new and powerful land development interests into Delta policy. In contrast to many water exporters, who have begun to question the viability of a major levee investment strategy, various in-Delta interests have stressed the importance of maintaining the integrity of the levee system.[5] At issue are both the salinity of water supplies and the viability of current land uses; both are at risk if the levees fail.

New Challenges for Water Supply Reliability

For water exporters, both ecosystem and levee issues have raised new questions about the ability to achieve the water supply reliability goals articulated under CALFED. These goals include protection from involuntary cutbacks in exports, increases in water use efficiency to reduce demand pressures, and increases in exports through improvements in conveyance and expanded water storage. From CALFED's inception, the expansion of exports has been the most contentious goal, with disagreements over the likely environmental consequences of new surface storage projects and the appropriate distribution of costs between water users and taxpayers for investments in such projects. As the investigation of new surface storage options languished in the first few years after the signing of the ROD, water exporters from the Central Valley Project and

---

[5]See, for instance, editorials in the *Stockton Record* (2005, 2006) and the *Contra Costa Times* (2006).

the State Water Project pushed ahead on proposals to increase exports through improvements in operations and conveyance systems.

The July 2003 "Napa Accord"—developed at a meeting of water project officials and contractors—set out a plan to enable pumping increases at the Tracy Pumping Plant under high inflow conditions. Although the process for developing the plan was highly contentious—given the absence of both fishery agencies and environmental groups from the bargaining table—it was eventually endorsed by CALFED management.[6] Relabeled the "South Delta Improvements Package," the plan now includes investments to maintain water levels and reduce water salinity in the southern Delta, in response to concerns of in-Delta interests (Cooper, 2003), with 3 to 5 percent greater average export volumes (mostly in high-flow years). However, by the time the environmental documentation for this package was available for public review in November 2005, the Delta's new ecosystem challenges had taken center stage, calling into question the feasibility of the plan's export enhancement goal.[7]

Meanwhile, the new spotlight on levee instability has focused exporters' attention on the reliability of the water conveyance system. Some of the most extensive public outreach efforts have been conducted by the Association of California Water Agencies (ACWA), whose "Blueprint for California Water," released in October 2005, calls on officials to "evaluate long-term threats to the Delta levee and conveyance system and pursue actions to reduce risks." As of this writing (October 2006), the Kern County Water Agency is the only exporting agency whose officials have publicly endorsed revisiting the peripheral canal (Associated Press, 2004). However, many exporters are concerned about the long-term viability of the Delta as a conduit. As Tim Quinn, vice president of the Metropolitan Water District of Southern California (MWDSC) noted, "The current policy of the state and that of our board is to move water through the

---

[6]On the disputes, see Pollard (2003) and Machado (2003). On the CALFED position, see CALFED (2004b) and Wright (2004).

[7]To wit, the DWR proposed to make decisions on the project in two separate stages, focusing first on the water level and quality and environmental objectives and only later on increasing exports (Department of Water Resources, 2005b). In a recent policy statement on the Delta, the Metropolitan Water District of Southern California (2006) emphasized only the water quality objectives of the project. On environmental community objections to the proposal, see Taugher (2006a).

delta. Mother Nature, however, has not been cooperating" (Lucas, 2005). In interviews, some water agency officials emphasized their concern that a strategy to shore up Delta levees would result in "stranded assets"— costing substantial investment dollars while leaving exporters vulnerable to curtailment of supplies.

Heightened Concern over Ecosystem Stress

Ecosystem stress has naturally been the primary concern for the environmental community. Given the history of battles to secure adequate environmental flows within the Delta, it is not surprising that many environmental groups looked to export levels as a likely culprit in the collapse of delta smelt and other pelagic species. In late 2005, Environmental Defense released a study reporting that environmental flows in the Delta had been considerably lower than targeted between 2002 and 2005, the period over which the fish decline set in (Rosekrans and Hayden, 2005). Even as scientific evidence has emerged suggesting that the decline is due to a more complex set of factors (Chapter 4), many within the environmental community remain convinced that export levels are at least partly to blame. In this, they have found allies among southern Delta farmers (Taugher, 2005).

In the late 1990s, a similar alliance between environmentalists and Delta farmers pulled a peripheral canal alternative off the table during the deliberations over the strategy to be pursued by CALFED. In light of new evidence on the Delta's woes, environmentalists have been divided over rethinking their position that the Delta must remain the only conduit for water exports. Gerald Meral, a Bay Area environmentalist and DWR official at the time of the original peripheral canal referendum in 1982, was one of the first to suggest that California reconsider such an option (Meral, 2005a, 2005b). Senator Joe Simitian, a Bay Area legislator with a strong environmental record, was the first to formally float a bill on this proposal (Taugher, 2006b). As various scientists, including those from the POD team, have indicated that such alternatives are worthy of consideration, some environmental groups have indicated a willingness to put them back on the table (Thompson, 2005a; Lucas, 2005; Gardner, 2006). Wariness remains, however, with some concerned that an alternative conduit for exports would lead to a decline in interest, resources, and commitment devoted to the Delta's ecological problems (Nelson, 2005).

Conflicts, Old and New

Recognition of the new threats to the Delta has reinforced long-standing conflicts over export levels, water quality, and ecosystem protection and has raised new conflicts and concerns over Delta land use. Increasingly, these conflicts are finding expression in legal actions.

Renewed Battles over Export Levels, Ecosystem Health, and Water Quality

Although legal actions had never entirely ceased during the decade-long CALFED truce, in 2005 a change in strategy took place on the part of environmental groups who had collaborated under CALFED. Various legal actions have been launched against federal and state agencies responsible for fisheries and water project management, on the grounds that they have favored water exports to the detriment of ecosystem health. Two lawsuits filed in 2005 challenged the biological opinions of federal regulatory agencies regarding the effects of new CVP operating criteria and plans (OCAP) on delta smelt and salmon.[8] In early 2006, the proposed CALFED intertie—or connector—between the CVP and SWP aqueducts, which would have increased export potential, was successfully delayed, sending project agencies back to the drawing board to complete more detailed environmental impact documentation.[9] Several groups petitioned the U.S. Fish and Wildlife Service to raise the delta smelt to endangered status under federal law, and in October 2006 a coalition of fishing groups sued DWR for failing to comply with state law protecting the smelt (Weiser, 2006a, 2006b).

---

[8] *Natural Resources Defense Council et al. v. Kempthorne et al.*, No. 1:05-CV-01207 OWW LJO (E.D. Cal. filed September 28, 2005) (delta smelt); *Pacific Coast Federation of Fishermen's Associations et al. v. Gutierrez et al.*, No. 1:06-CV-00245 OWW LJO (E.D. Cal. filed January 24, 2006) (salmon). Filing dates reflect when the cases were transferred from the Northern District to the Eastern District.

[9] *Planning and Conservation League v. U.S. Bureau of Reclamation* (C05-3527 CW, N.D. Cal., filed February 15, 2006).

Over this period, decisions on several legal and regulatory actions added to the mounting challenges against water exports. In October 2005, a state appeals court ruled that parts of the CALFED environmental impact review were inadequate, notably because the review had failed to consider the option of reducing exports (Boxall, 2005). The following February, the State Water Resources Control Board issued a cease-and-desist order against the CVP and the SWP, threatening to cut back pumping levels if the agencies failed to implement a plan to maintain salinity standards for agriculture in the southern Delta (Barbassa, 2006; State Water Resources Control Board, 2006). During the spring and summer of 2006, under the threat of a court-mandated reallocation of project water, water users and environmentalists negotiated a settlement to a decade-old lawsuit to restore environmental flows to the San Joaquin River.[10] In April, the National Marine Fisheries Service announced the listing of yet another species that migrates through the Delta, the southern green sturgeon, as threatened under the federal Endangered Species Act (ESA). In July, responding to one of the OCAP lawsuits, the U.S. Bureau of Reclamation requested a reexamination of the effects of Delta export pumping on the delta smelt (Young, 2006).

Exporters, meanwhile, have been pursuing the creation of a Habitat Conservation Plan (HCP) as an alternative approach to CALFED for ecosystem issues in the Delta. Instead of relying on biological opinions of the fisheries agencies to determine ESA regulatory actions (such as the timing and volumes of water exports), an HCP would authorize interested parties to develop and invest in a long-term, multispecies protection plan. These parties would then receive ESA coverage (i.e., permission for some "takings"—or deaths—of listed species) for a range of activities. Exporters see this approach—and its California law counterpart, the Natural Communities Conservation Plan (NCCP)—as more flexible and likely to succeed than the approach used to date.[11] The state and federal fisheries agencies and several environmental groups have endorsed this process,

---

[10] *Natural Resource Defense Council et al., v. Rodgers et al.*, Stipulation of Settlement, CIV No. S-88-1658-LKK/GGH (filed at the U.S. District Court of Sacramento on September 13, 2006).

[11] See the University of the Pacific "Statement of Principles" (anonymous, 2005).

known as the Bay-Delta Conservation Plan (BDCP). Exporters hope to involve other actors whose behavior affects ecosystem health, including power generators at the Delta's western edge and upstream operators and diverters. To date, the BDCP's scope is more limited than those developed in various parts of Southern California (notably, Riverside and Orange Counties), in which local land use authorities (cities and counties) are active participants. As discussed further in Chapter 9, the omission of land use interests will limit the BDCP's potential to play a coordinating role, given the central role of land use decisions—and particularly urbanization—in the management of Delta environmental resources.

New Conflicts over Land Development

The 1992 Delta Protection Act had aimed to set limits on urbanization by designating the lowest and most subsided islands as a "primary zone," reserved principally for agricultural, environmental, and recreational use (Figure 3.5). The act did not attempt to regulate development in the "secondary zone"—consisting of upland areas as well as some low-lying lands already zoned for development. From the act's passage until the failure of Lower Jones Tract levee in the summer of 2004, land development in the Delta had maintained a relatively low profile, with urbanization plans proceeding in the secondary zone. This changed with increased recognition of flood risks in the Delta, particularly in the aftermath of Hurricane Katrina.

Urbanization in the Delta is an issue on which most other stakeholders—including some Delta farmers—are able to agree: They think it is a bad idea (Machado, 2005; Pitzer, 2006). The concerns not only include increased risks of economic damage and threat to human life from floods in Delta lowlands, they also include potential threats to water quality and a loss of wildlife habitat. As Chapter 4 points out, the "hardening" of Delta uplands is also relevant for long-term wildlife habitat options, given the likelihood of eventual flooding of many Delta islands. On the other side of this issue are developers and local land use authorities, as well as some farmers hoping to sell their land at high prices. The issue is not strictly one of profits. For local authorities, new development is often seen as a way to increase tax revenues and finance improvements in local infrastructure, including better flood protection for existing residences.[12]

The first signs of a formal challenge to the 1992 partition of Delta lands emerged in the spring of 2006, when environmental groups filed legal actions against two developments within the secondary zone. One lawsuit sought to block a 4,000 home project on Hotchkiss Tract, which lies below sea level (Hoge, 2006a). The suit argued that the City of Oakley had failed to consider adequately the risks of levee failure or mitigation of the likely effects of urban development. A second action challenged the state Reclamation Board's decision to approve a developer's levee-widening proposal on Stewart Tract, now part of the City of Lathrop (Hoge, 2006b). Recalling that this island, which lies above sea level, lay under 10 feet of water during the 1997 floods, the appeal challenged that its development would "exacerbate and worsen the existing flood threat for current and future residents."

The Context for a New Delta Vision

In several respects, the current situation is reminiscent of the turmoil in the years preceding the 1994 Bay-Delta Accord, with serious concerns over ecosystem health and a rise in legal and regulatory actions that threaten to curtail water exports. Now, as then, various interests with a stake in the Delta are embarking on an exercise to seek a new course of action. The task for the governor's Delta Vision effort—today's equivalent of the CALFED process—is even more complex. On the one hand, new stakeholders have emerged—notably developers and Delta cities promoting urbanization of Delta lowlands—with even stronger interests in maintaining parts of the Delta in their current form. On the other hand, new scientific analysis— described in Chapter 3 of this report—has shown that this goal may not be viable, given the various pressures on the levee system. Moreover, as described in Chapter 4, maintaining the current configuration of Delta water flows may not be in the best interests of the fish species that are now under threat. There is also less promise of state and federal funds to

---

[12]See, for instance, the commentary by the city manager of Oakley (Montgomery, 2006).

lubricate and finance any agreement. For these reasons, a new agreement based on the maxim that "everyone gets better together" may be elusive, because some goals for the Delta inherently conflict.

Three questions will inevitably be central to any process to forge agreement among stakeholders on a new Delta vision. First, what capacity is there to adjust to changing conditions in the Delta? Recognition of adjustment capacity opens up the possibility to consider a wider range of options for the Delta's future. Some stakeholders are already taking steps to reduce their exposure to risk from levee failure. Among water exporters, the Southern California agencies belonging to MWDSC's vast network are probably furthest along this path. Investments over the past decade in water marketing contracts, groundwater and local surface storage, conservation, recycling, and other local resources have put the region in a position to ride out an outage of Delta water supplies for up to two years.[13] Water agencies in the Bay Area are increasing their resiliency through investments in conservation and recycling, interties, and plans for regional desalination facilities. Such adjustments are not limited to water exporters. For instance, PG&E is laying a new pressurized gas pipeline underground to reduce its vulnerability to island flooding. And although there is disagreement over their adequacy to mitigate flood risks, some Delta land developers have proposed larger levees than the legal minimum.

The second question is how will California pay for any given set of options, be it shoring up the existing levees, building a peripheral canal, or any other substantial alternative? A divergence in views has already emerged. Various interests within the Delta have hinted that water exporters should foot the lion's share of the bill, given the importance of the Delta as a conveyance system. Exporters, meanwhile, are emphasizing their unwillingness to pay more than their "fair share," along with all the other Delta interests.[14] Failure to agree on workable principles for applying a beneficiary pays criterion to Delta investments puts any new visioning exercise at risk of coming up short, as did CALFED. As discussed further

---

[13] See, for instance, the comments by MWDSC general manager Jeffrey Kightlinger (Pitzer, 2006).

[14] Metropolitan Water District of Southern California (2006) and California Urban Water Agencies (2006).

in Chapter 9, the state bond funds approved for flood control in November 2006 would provide only a down payment on any long-term strategy.

The third question is how will the various legal actions now under way or planned interact with more consultative processes? Threat of legal and regulatory actions brought some water users to the table in the early 1990s, and this is certainly still a way to force compromise on issues relating to environmental protection. However, there is also a risk that court rulings will constrain the consideration of new alternatives for the Delta, because so much of the focus of the lawsuits has been on limiting exports while maintaining the Delta as a levee-dependent freshwater body.

In the remaining chapters of this report, we explore some of these issues in greater detail. Chapter 6 assesses the capacity of various water users—including exporters and those who draw indirectly from the Delta—to adjust to changes in volumes and salinity levels. Chapters 7 and 8 examine a wide set of options for the Delta's future and evaluate the ability of these alternatives to "deliver" with respect to various Delta goals. Chapter 9 looks at questions of financing and governance, with a particular focus on how to mitigate the costs for those who would bear disproportionate adjustment burdens, and considers possible policy realignments for a new Delta vision.

Although the current crisis has similarities with the previous debates—on the peripheral canal in the 1970s and early 1980s and during the initial CALFED discussions in the early 1990s—there are now new interests and concerns. It is important to move the policy discussion beyond the choice between a levee-centric freshwater Delta and a peripheral canal.

6. Water Supply Adaptations to Changes in Delta Management

"The rains of California are ample, but confined to Winter and Spring. In time, her streams will be largely retained in her mountains by dams and reservoirs, and, instead of descending in floods to overwhelm and devastate, will be gradually drawn away throughout the Summer to irrigate and refresh. For a while, water will be applied too profusely, and injury thus be done; but experience will correct this error; and then California's valleys and lower slopes will produce more food to nourish and fruit to solace the heart of man than any other Twenty Millions of acres on earth."

Horace Greeley (1868), Recollections of a Busy Life

In this chapter, we examine how water users in California might adapt to major changes in the Delta and in Delta water management. Water agencies and users have a wide range of long-term options in this regard. The exploration and integration of these options in complex water systems usually require the use of computer models. Here, we employ two computer models of water and agricultural management to examine adaptations and adaptation costs for several major, even extreme, sets of long-term Delta conditions. The CALVIN (California Value Integrated Network) model examines long-term statewide water supply adaptations to changes in Delta water availability. The DAP (Delta Agricultural Production) model examines how changes in Delta salinity might affect agricultural production within the Delta. We also briefly review the benefits of a peripheral canal water supply diversion upstream of the Delta and consider its economic value, based on the results of our modeling exercises. This analysis provides useful background for a broader discussion of alternatives for the Delta, pursued in the next chapter. We begin with a review of the direct and indirect use of water from the Delta in different parts of the state.

State and Regional Use of Delta Water Supplies

Table 6.1 presents estimates of the consumptive uses of water (water that is either consumed or evaporated and unavailable for potential reuse) in or tributary to the Sacramento–San Joaquin Delta. Because these estimates must be assembled from various sources, the particular numbers are somewhat uncertain. Nevertheless, they illustrate some important points.

First, there is little doubt that much less water flows through the Delta today than would under natural conditions.[1] In an average water year (October to September), total diversions from the Delta—about 18 million acre-feet (maf)—account for roughly 40 percent of all flows that would have naturally passed through the Delta. In addition, the seasonal patterns of Delta inflows and net outflows have been altered significantly. Today, spring Delta outflows are much lower than they would be naturally, and summer outflows are generally higher.

Second, most diversions (64% on average) occur upstream of the Delta. To the north, Sacramento Valley water users deplete Delta inflows by almost 6.7 maf per year, mostly for agricultural uses. To the south, an additional 4.0 maf per year are consumed by diversions on the San Joaquin River and its tributaries, including the Friant-Kern Canal, which exports water to the Tulare Basin (Kern and Tulare Counties). The major water projects that use the Delta as a transfer point—the Central Valley Project and the State Water Project—account for only about 31 percent of all diversions, averaging 5.4 maf per year and regularly exceeding 6.0 maf per year in recent years. The balance (4%) is accounted for by in-Delta users, primarily farmers.

Third, direct exports from the Delta have increased over time, with the exception of drought periods (Figure 6.1). This trend continues today. Although exports to the federal Central Valley Project have decreased somewhat in recent years as a result of environmental flow requirements of the CVPIA, State Water Project exports have increased in response both to growth in urban water demand in Southern California and the Bay Area and to several recent wet years.

---

[1]There is some dispute over the extent to which native vegetation and wetlands consumed some of these flows under natural conditions. Also, precipitation increases in recent decades might be mitigating some effects of increased water withdrawals (Fox, Mongan, and Miller, 1990).

Table 6.1

Estimated Average Consumptive Uses of Delta and Delta Tributary Waters, 1995–2005 (taf/year)

Demand Area	Agriculture	Urban	Environment[a]	Total
Net Delta outflow	—	—	22,553	22,553
Total diversions	14,090	3,235	415	17,740
Upstream diversions	9,540	1,712	138	11,390
Delta diversions	4,550	1,523	277	6,350
In-Delta	769	0	0	769
Upstream diversions	0	0	0	0
Delta diversions	769	—	—	769
North of Delta	6,000	562	138	6,700
Upstream diversions	6,000	520	138	6,658
Delta diversions	0	42	0	42
South of Delta	7,321	1,699	277	9,297
Upstream diversions	3,540	600	—	4,140
Delta diversions	3,781	1,099	277	5,157
West of Delta	0	974	0	974
Upstream diversions	0	592	0	592
Delta diversions	0	382	0	382

SOURCES: U.S. Bureau of Reclamation (2005); Jenkins et al. (2001), Appendix F; Department of Water Resources (1998, 2005c); DAYFLOW data (Department of Water Resources); San Francisco Public Utilities Commission (2005); Santa Clara Valley Water District (2005); Contra Costa Water District (CCWD) (2005); and East Bay Municipal Utilities District (2005).

NOTES: Calculations assume that consumptive use constitutes 75 percent of upstream agricultural withdrawals and 65 percent of upstream urban withdrawals. taf = thousand acre-feet.

[a]Environmental uses include net Delta outflows and water diverted to supply wetlands.

Given anticipated population growth over the coming decades, California's urban water demand is likely to increase, although conservation programs will slow the pace of this growth. However, agricultural water uses are likely to decline somewhat in reaction to market forces, including land development (Department of Water Resources, 2005c). Some agricultural lands south of the Delta also will be coming out of production because the soils are becoming too saline to farm profitably. Some growth

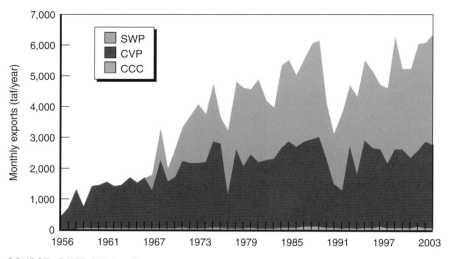

SOURCE: DAYFLOW data (Department of Water Resources).

NOTES: Totals are for water years (October to September). Exports include the Central Valley Project at Tracy (Delta-Mendota Canal), the State Water Project at Banks (California Aqueduct), and diversions for the Contra Costa Water District through the Contra Costa Canal (CCC). Five-year gap between first two years, six-year gap between others.

Figure 6.1—Major Direct Water Exports from the Delta, 1956–2005

in urban water demands can be offset by these declines in irrigation, as well as by improvements in water conservation. On balance, only small increases in total water demands are likely for urban and agricultural uses.

Delta water supplies remain highly variable, despite substantial management of flows through reservoir storage and releases. Inflows to the Delta from upstream sources vary greatly across seasons and years (Figure 6.2). The driest year of record (1976–1977) had little inflow, averaging only 2,800 cfs for the year, and little absolute seasonal variability, ranging from 1,600 to 5,000 cfs. The wettest year of record (1982–1983) had an average inflow of 89,000 cfs, ranging from 23,000 to 267,000 cfs of monthly average flows. Other years of record had higher individual monthly flows, usually associated with floods. We estimate that on average, the inflows that would have occurred if the Delta had been in its natural

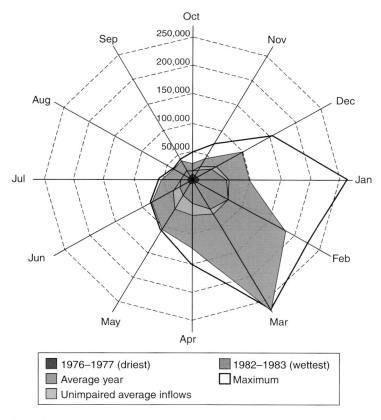

Oct
250,000
Sep
Nov
200,000
Aug
150,000
Dec
100,000
50,000
Jul
Jan
Jun
Feb
May
Mar
Apr

■ 1976–1977 (driest)   ■ 1982–1983 (wettest)
■ Average year   □ Maximum
■ Unimpaired average inflows

SOURCE: DAYFLOW data (Department of Water Resources).

**Figure 6.2—Seasonal and Annual Variability of Delta Inflows,
1956–2005 (cfs)**

state (shown as "unimpaired" flows in Figure 6.2) tended to be greater than current inflows, especially during spring.[2]

Direct water exports from the Delta are also variable (Figure 6.3), although to a lesser extent than inflows. There are two distinct seasons of

---

[2]Unimpaired flows are estimated using two DWR data series for the period 1956–2005: (1) DAYFLOW estimates of Delta inflows and exports and (2) estimates of unimpaired or natural Central Valley inflows.

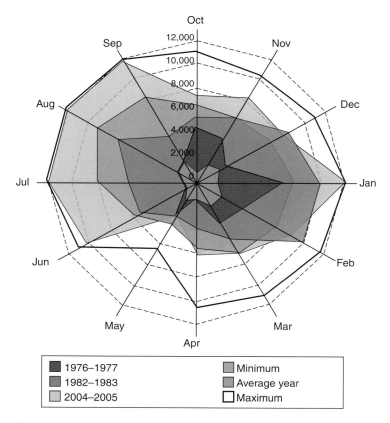

Figure 6.3—Seasonal and Annual Variability of Delta Pumping, 1975–2005 (cfs)

SOURCE: DAYFLOW data (Department of Water Resources).

pumping, winter and summer, with historically less pumping in spring and fall months. This pattern is a result of the high demand for irrigation water during the summer months and the filling of off-stream storage in San Luis Reservoir in winter. It also reflects efforts to minimize pumping during the spring and fall months when fish are spawning. Annual export pumping since 1975 has ranged from 3,100 cfs (1976–1977) to 8,900 cfs (2004–2005).

Statewide Adaptations to Delta Water Availability and Management

The reliability of the Delta as a water source is of great concern to water managers, particularly those whose agencies rely on direct diversions of Delta water. At issue are both regulatory reliability (given continued concerns over the needs of fish species) and physical reliability (given the threats to the integrity of the levee system). In response, many users of exported water have made strides to reduce their dependence on the Delta in recent years. Urban water agencies have been developing interties—or connectors between aqueducts—to enable water sharing in the event of emergencies, such as a massive levee failure. Both urban and agricultural water agencies have developed underground storage (or "groundwater banking"), water use efficiency, water markets, transfers and exchanges, wastewater reuse, and other activities. Indeed, much can be done to reduce the dependence of water users on Delta supplies, although such actions always come at some cost, in terms of financial expense or water scarcity (i.e., using less than the desired amount of water).

If water supplies from the Delta were abruptly cut off and water users were both unable to draw on alternative supplies and unprepared to reduce water use, the results would be catastrophic for many users. Costs for such scenarios, arising from multiple levee failures, are estimated to be as high as $10 billion per year (Illingworth, Mann, and Hatchet, 2005). In contrast to these scenarios, this chapter examines a "soft landing" approach to adaptation, in which reasonable preparations would be made for any major changes in Delta conditions and management.

Water suppliers and users can be remarkably adaptable. Studies of how California's water supply could adapt to major climate, population, and infrastructure changes indicate that considerable adjustment is physically possible at reasonable cost (Tanaka et al., 2006; Jenkins et al., 2004). Furthermore, adaptations may be facilitated by the highly intertied nature of the state's water system and the decentralized nature of water management. State and federal agencies manage the large water projects, but many planning decisions are made at the local level. Local and regional water agencies commonly have the political, financial, and technological wherewithal to make long-term changes in their water supplies and water use. Although institutional conflicts can limit short-term actions, cooperation has increased considerably in recent decades in such areas as water marketing, groundwater banking, and emergency sharing agreements.

Table 6.2 summarizes many of the options available to water managers seeking to balance supplies and demands. In addition to traditional methods to expand usable water supplies, such as surface storage, conveyance, and water treatment, the list includes more contemporary methods, such as improvements in operational efficiencies and wastewater reuse. Water demand management measures include improving water use efficiency ("more crop per drop") as well as water scarcity (reducing water use beyond desired levels by rationing urban water use, fallowing some farmland, or curtailing recreational activities). Various general tools (pricing, water markets, exchanges, and taxes or subsidies) may be used to motivate local users to implement both supply- and demand-side options.

Modeling Water Supply and User Adaptations

A similarly wide range of alternatives exist for managing Delta water supplies. As seen in Chapter 2, numerous alternatives have been proposed in the past, and Chapter 7 will consider others. Various Delta outflow regulations, policies on Delta exports, changes in physical pumping, conveyance, and storage capacities would be reasonable elements to examine, both individually and in combinations. If one also considers a reasonable set of adaptations by water users and managers, estimating the performance of alternatives becomes a complex exercise Here, we draw on two computer models to examine the ability of California water users to adapt to changes in water supply available from the Delta. The CALVIN model explores how California's larger water supply system could respond to changes in water supplies and demands resulting from different Delta management strategies. The DAP model explores how in-Delta agriculture would be affected and would respond to changes in Delta land and water management.

All model results are based on imperfect assumptions and limited information. Nevertheless, for such complex systems as the Delta and California's water supply, these types of analytical aids are indispensable for exploring, developing, and evaluating new alternatives. Computer models

Table 6.2

Water Supply System Management Options

Demand and Allocation Options

General

Pricing[a]

Subsidies, taxes

Regulations (water management, water quality, contract authority, rationing, etc.)

Water transfers and exchanges (within or between regions/sectors)[a]

Insurance (drought insurance)

Demand Sector Options

Urban water use efficiency[a]

Urban water scarcity[a] (water use below desired quantities)

Agricultural water use efficiency[a]

Agricultural water scarcity[a]

Ecosystem restoration/improvements (dedicated flow and nonflow options)

Ecosystem water use effectiveness

Environmental water scarcity

Recreation water use efficiency

Recreation improvements

Recreation water scarcity

Supply Management Options

Operations Options (Water Quantity or Quality)

Surface water storage facilities (new or expanded)[a]

Conveyance facilities (new or expanded)[a]

Conveyance and distribution facility operations[a]

Cooperative operation of surface facilities[a]

Conjunctive use of surface and ground waters[a]

Groundwater storage, recharge, and pumping facilities[a]

Supply Expansion Options (Water Quantity or Quality)

Supply expansions through operations options (reduced losses and spills)

Agricultural drainage management

Urban water reuse (treated)[a]

Water treatment (surface water, groundwater, seawater, brackish water, contaminated water)[a]

Desalting (brackish and seawater)[a]

Urban runoff/stormwater collection and reuse (in some areas)

[a]Options represented in the CALVIN model.

allow us to precisely represent current knowledge and explore the implications of uncertainties in a standardized evaluation of a wide range of solution alternatives. Although there are obvious pitfalls to quantitatively analyzing such complex systems, decisionmaking without such aids has shown itself to be risky, even dangerous. Model results provide insights based on our best knowledge of the system and a relatively transparent way to compare policy and management alternatives for complex systems.

Delta Agricultural Production Model

The DAP model specifically focuses on agricultural land, water, and cropping decisions for the Delta region. It is calibrated on four years of recent agricultural land use data for Delta lands and crop production as well as on cost data for nearby regions. DAP allows cropping and water use decisions, and their associated revenues and profits, to be estimated for 35 Delta subregions (individual islands and groups of islands) for a set of salinity conditions (see Appendix D). We use the DAP model below to examine how Delta cropping patterns and profitability might change under Delta management alternatives that alter salinity.

Statewide Economic-Engineering Water Supply Model (CALVIN)

The CALVIN economic-engineering optimization model represents California's vast intertied water supply and demand system. The model was developed with state funds over the past eight years (Jenkins et al., 2001; Draper et al., 2003) and has been applied to a variety of water management problems, including problems of climate change (Jenkins et al., 2004; Pulido-Velázquez, Jenkins, and Lund, 2004; Null and Lund, 2006; Tanaka et al., 2006). This model is used below to examine the effects of changes in water exports on all major agricultural and urban water users that depend on the Delta. The model includes a wide range of adaptation options (Table 6.2). The scenarios are based on water demand for the year 2050, with a projected state population of 65 million (up from 37 million in 2005). They also assume that water agencies will complete currently planned infrastructure enhancements. Although fixed and construction costs are not included, the modeled results put the water supply costs and responses of each management alternative in perspective. Appendix C contains more detailed information on the model as well as additional model results for the cases discussed here. CALVIN is intended as a strategic screening model to identify promising operations and management alternatives and to provide preliminary water supply cost estimates of these alternatives.

Delta Management Alternatives

We focus on three Delta management alternatives that illustrate the water management and performance implications of a wide range of Delta policies and show how modeling analysis can provide insights for crafting and evaluating Delta alternatives. The three alternatives are an abandonment of water exports, a substantial increase in minimum net outflow requirements from the Delta into the San Francisco Bay, and a shift to allowing parts of the Delta to become more saline. We employ the CALVIN model to better understand major water supply consequences and potential water management responses to the first two alternatives, and we employ the DAP model to assess the consequences to Delta agriculture of the third alternative.

Effects of Ending Delta Exports

An extreme policy alternative would be to completely abandon all exports from the Delta. Although extreme, such an alternative could be imagined if the Delta proved to be an excessively unreliable or expensive part of California's water supply system. To model responses without Delta exports, we assumed that this change is not sudden, as might occur in the case of an unforeseen, catastrophic levee failure. Rather, we assume that water agencies would become well prepared for the change, by constructing reasonable interties, wastewater reuse, and desalination facilities and fashioning institutional agreements to cooperate, such as water marketing and exchanges.

Economic and water delivery results under 2050 demand conditions appear in Table 6.3. "Target delivery" refers to estimates of the annual water deliveries that would eliminate shortages for each water service area, irrespective of costs. "Delta exports" assume cost-effective (optimized) operations with current levels of access to Delta pumping. This assumption results in an average 2.9 maf per year of shortages (as indicated by the "water scarcity" column, which shows "target" minus "delivery"). "Scarcity cost" is the economic cost to local water users of these shortages. This includes lost agricultural production and the costs to households

Table 6.3

Average Annual Water Scarcity Costs Without Delta Exports

Region	Target Delivery (taf)	Delta Exports			No Exports			Net Change	
		Delivery (taf)	Water Scarcity (taf)	Scarcity Cost ($ million)	Delivery (taf)	Water Scarcity (taf)	Scarcity Cost ($ million)	Delivery (taf)	Scarcity Cost ($ million)
Statewide	42,016	39,066	2,950	210	34,055	7,961	1,041	–5,011	831
Urban	12,809	12,749	60	44	12,461	347	321	–287	277
Agriculture	29,208	26,317	2,891	166	21,594	7,614	720	–4,723	554
Sacramento Valley	10,935	10,617	318	3	10,798	137	1	181	–2
Urban	1,662	1,662	0	0	1,662	0	0	0	0
Agriculture	9,274	8,956	318	3	9,137	137	1	181	–2
San Joaquin Valley	8,025	7,405	620	10	6,129	1,896	138	–1,276	128
Urban	1,634	1,634	0	0	1,605	29	34	–29	34
Agriculture	6,391	5,771	620	10	4,524	1,866	104	–1,247	94
Tulare Basin	11,679	10,667	1,012	24	7,010	4,669	486	–3,657	462
Urban	1,406	1,406	0	0	1,406	0	0	0	0
Agriculture	10,272	9,260	1,012	24	5,603	4,669	486	–3,657	462

Table 6.3 (continued)

Region	Target Delivery (taf)	Delta Exports			No Exports			Net Change	
		Delivery (taf)	Water Scarcity (taf)	Scarcity Cost ($ million)	Delivery (taf)	Water Scarcity (taf)	Scarcity Cost ($ million)	Delivery (taf)	Scarcity Cost ($ million)
Southern California	11,378	10,377	1,001	173	10,118	1,260	415	–259	242
Urban	8,107	8,047	60	44	7,788	318	286	–259	242
Agriculture	3,271	2,330	941	129	2,330	942	129	0	0

SOURCE: CALVIN model.

NOTES: The table reports results for projected water demands in the year 2050. Regional numbers might not sum to the statewide total because of rounding. Bay Area urban users and Delta agriculture are included in the San Joaquin Valley, and Central Coast urban contractors of the SWP are included in the Tulare Basin.

and industries of water conservation and other reductions in use. For the first case, statewide scarcity cost averages $210 million per year.

The no-export case precludes all Delta exports from the CVP, the SWP (except the small North Bay Aqueduct), and the CCWD. Infrastructure is the same as the first case, except that additional intertie capacity would be constructed, mostly where some aqueducts currently cross or are nearby (Appendix C).

Sectoral and Regional Effects

The CALVIN analysis demonstrates that California's economy as a whole would not suffer catastrophic consequences if direct Delta exports were ended in a well-planned manner. Without water exports, annual costs to water users would be on the order of $831 million, less than one-tenth of one percent of the state's current $1.5 trillion per year economy. This contrasts with much higher costs (on the order of $10 billion per year) if Delta exports were ended abruptly (Illingworth, Mann, and Hatchet, 2005). But even under a well-planned abandonment of Delta exports, the economic costs to water importing regions of the state would be substantial, including roughly $554 million per year in reduced agricultural production and $277 million per year in increased water scarcity for urban areas. Overall, water deliveries would fall by five maf per year, and the brunt of this loss would be felt by agricultural water users within and south of the Delta (in the San Joaquin and Tulare regions), who would lose about a third of their deliveries.[3]

With so many changes in water supply deliveries and operations, operating costs for the water supply system would also change significantly. These costs would include pumping, water and wastewater treatment, and costs for additional wastewater reuse (at $1,000 per acre-foot) and seawater desalination (at $1,400 per acre-foot).[4] The costs of expanding wastewater

---

[3]For details, see Appendix Figures C.2 through C.5. Such large reductions in output might also raise the price of some commodities, particularly those for which San Joaquin Valley farmers have a large market share, such as some fruits and vegetables. This shift would augment revenues for farmers who can remain in production (in California or elsewhere) and generate some additional costs for consumers (in California and elsewhere).

[4]These are conservative cost estimates for these new sources. The most recent California Water Plan Update (Department of Water Resources, 2005c) assumed a range

reuse and desalination (over $1 billion per year) would be largely offset by reductions in pumping and treatment costs for export-based water deliveries, leaving an overall increase in operating costs of only $157 million (Table 6.4). Thus, the overall water scarcity and operating cost of ending direct Delta exports would be about $1 billion per year.

Table 6.4

Average Annual Operating Costs Without Delta Exports ($ million)

	Delta Exports	No Exports	Cost Increase
Statewide	3,154	3,311	157
Sacramento Valley	195	206	12
San Joaquin Valley and Tulare Basin	998	974	−24
Southern California	1,961	2,131	169

SOURCE: CALVIN model results for water demands in the year 2050.

Without exports, urban areas that rely on Delta exports would initially lose important supplies, but they would be able to compensate with various alternative sources. Water deliveries from the Delta to urban Southern California would decrease by about 2.2 maf per year, but water purchases and recycling investments would reduce this gap nearly tenfold, to 258 taf per year. Urban water users in the Bay Area would be able to adapt with increased intertie capacity, more wastewater reuse, and seawater desalination. Central Coast cities supplied by the SWP would also be shorted, and they would need to increase wastewater reuse and seawater desalination.

Agricultural water users south of the Delta would also make considerable use of water markets, conjunctive use, and increases in water use efficiency. However, the net water delivery and economic effects would still be substantial, particularly for farmers on the west side of the San Joaquin Valley and Tulare Basin, who depend most on Delta pumping. Agricultural areas dependent on San Joaquin River diversions at Friant Dam and Tulare Basin inflows would also be affected, because these would remain the

---

of $300 to $1,300 per acre-foot for recycled water and $800 to $2,000 per acre-foot for seawater desalination.

only transportable surface waters that could serve regions whose exports have been cut off. Tulare Basin agricultural production would be particularly affected by the end of Delta water exports, although many farmers with rights to Friant-Kern and local Tulare surface waters would be likely to do well financially through sales of this scarce water to cities in Southern California.

Meanwhile, other agricultural areas in the state would be largely unaffected by ending water exports from the Delta. Agricultural areas on the east side of the San Joaquin Valley that depend directly on streams flowing from the Sierra Nevada would be much less affected, because they do not depend on the Delta and they cannot transfer water to other regions without sending water through the Delta. Inland Southern California agricultural users, who rely predominantly on Colorado River water supplies, would be unaffected because the Colorado River Aqueduct has no available capacity to transport additional water to Southern California cities. (The "Delta exports" case assumes that enough transfers would have already taken place to keep this aqueduct full.) The end of Delta exports would cut Sacramento Valley farmers off from transfer opportunities; instead, their water deliveries and agricultural profits would increase slightly because they would no longer need to contribute to Delta outflows. Sacramento Valley cities would be unaffected.

The end of direct Delta exports would reduce some pressure on environmental flows in the Sacramento Valley and Trinity River. However, wetland water deliveries south of the Delta would become much more expensive in terms of additional scarcity costs to other uses.[5]

Storage Versus Conveyance

Without Delta exports, the value of water storage capacity would decrease in most locations. South of the Delta, surface water storage sites would tend to be emptier because there would be less water to keep in storage. North of the Delta, reservoirs would tend to have more water but would no longer be able to help alleviate water problems in the southern part of the state. The only exceptions would be modest increases in the value of storage capacity at Millerton on the San Joaquin River and

---

[5]For details, see Appendix Table C.3.

in reservoirs on inflows to the Tulare Basin (especially on the Kaweah and Tule Rivers). For no reservoir would the average economic value of increasing storage capacity exceed $100 per acre-foot per year.[6]

Instead, conveyance capacity would become much more valuable, reflecting the value of moving available water sources to places that lose export supplies.[7] For instance, the average economic value of expanding the Hayward-EBMUD intertie would increase from $178 per acre-foot to $588 per acre-foot. The value of expanding the Hetch Hetchy Aqueduct would rise to $608 per acre-foot. Expansion of both facilities would replace some lost State Water Project supplies. Expansion of Colorado River Aqueduct capacity would rise in value from $169 per acre-foot to $488 per acre-foot, reflecting increased water scarcity and operating costs in Southern California. Capacities along the Cross Valley Canal in the Tulare Basin would also merit consideration for expansion, with economic values averaging $151 per acre-foot. This expansion would allow more San Joaquin and Tulare Basin inflows to be diverted to the California Aqueduct for Southern California water users. The value of increasing Mokelumne River Aqueduct capacity, to allow greater diversions from the Mokelumne River or the Sacramento River (through the Freeport Project), would average $186 per acre-foot. The value of a small peripheral canal—allowing continued exports of Northern California water—would be roughly the same.

Effects of Climate Change

With climate warming, the costs of eliminating Delta exports could increase substantially. This increase could arise in two ways. First, decreases in precipitation—predicted by some climate models—may reduce overall water availability. Second, the diminished storage capacity of the Sierra Nevada snowpack—foreseen by all climate models—will reduce the ability to move water from surplus times and locations (winter in Northern California) to surface and groundwater storage locations elsewhere in the

---

[6]In other words, users would not be willing to pay more than $100 per acre-foot for additional storage—a lower value than the per acre-foot cost of most, if not all, surface storage programs. See also Appendix Table C.4.

[7]The costs of such conveyance facilities are not available and would vary greatly with local conditions, but are commonly $1 million to $3 million per mile of length.

state. Available climate warming adaptation studies indicate that these conditions would increase the value of using direct Delta exports to move water from wetter to drier seasons and locations (Tanaka et al., 2006; Lund et al., 2003; Medellin et al., 2006). Therefore, the loss of Delta exports could constitute a more significant loss to the state as the climate changes over time.

Soft Versus Hard Landings

Even with tremendous preparation and forethought, ending all exports from the Delta would have substantial regional economic effects on California, averaging $1 billion per year in increased water scarcity and operating costs. Although this is a large effect, it is much smaller than the economic consequences of a sudden loss of the Delta because of catastrophic levee failure, an effect estimated at up to $10 billion per year. However, a series of infrequent Delta catastrophes, or hard landings, each entailing Delta failure, severe shortage, and rebuilding, might be less expensive overall than the permanent ending of all exports. In any event, either a series of hard landings or the ending of direct Delta exports would have very substantial and probably unacceptably high economic and political costs. However, the development of a soft landing strategy will require state and local leadership and preparation, as well as the negotiation of major changes in institutions, regulations, contracts, and finance (see Chapter 9).

Effects of Increasing Minimum Delta Outflow Requirements

Allowing greater levels of net Delta outflows into the San Francisco Bay is the traditional method for reducing seawater salinity in the Delta. It is not surprising, therefore, that those interested in preserving the Delta as a freshwater body—including Delta farmers, local urban diverters, and some environmentalists—often call for increases in net Delta outflows. This objective might gain more support in light of concerns over levee stability. If many island levees fail, or if the sea level rises substantially, increased Delta outflows might be needed to maintain the freshness of the western Delta. Because increasing net Delta outflows reduces the amount of water available for direct water exports and upstream diversions, it poses a threat to many water users, particularly south of the Delta.

We used the CALVIN model to examine the effects of increases in minimum Delta outflow requirements on water operations in California. Although this strategy has some similarities to the prohibition of Delta exports, the water operations and economic consequences are considerably different. Whereas export prohibition effectively excludes upstream diverters in the Sacramento Valley and some eastside San Joaquin Valley communities from participating in adjustments (because they have no way to send their water to exporters if Delta exports are prohibited), increases in minimum Delta outflows allow all regions that use Delta water to participate in adaptations.

The most cost-effective way to increase net Delta outflows would use a dual strategy that reduces both upstream water diversions and direct exports from Delta pumping plants (Figure 6.4). Assuming that the regulatory burden for these reductions would fall on export water users south of the Delta, who have lower priority water rights, this strategy would require a substantial increase in water sales moving through the Delta. For example, these water users, including urban agencies and farmers in the western San Joaquin Valley and the Tulare Basin, would be willing to pay Sacramento Valley and eastside San Joaquin water users to reduce their own use and allow more water to flow into the Delta via the Sacramento and San Joaquin Rivers; from there, much of it would be pumped south of the Delta to urban agencies and farms in the western San Joaquin Valley and the Tulare Basin. Water users would also make greater use of wastewater reuse, cooperative operations, and water conservation. In contrast to the no-export scenario examined above, seawater desalination would be used only in extreme circumstances.[8]

The economic cost of water scarcity for agricultural and urban sectors in each region and overall is shown in Figure 6.5. In contrast to the earlier no-export case, a strategy of increasing net Delta outflows would mean that burdens and incentives for cost-effective water management were spread

---

[8]For details, see Appendix Figures C.8 to C.10.

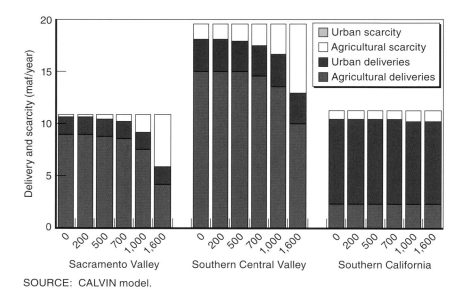

SOURCE: CALVIN model.

Figure 6.4—Average Agricultural Water Scarcity by Region with Increasing Minimum Monthly Net Delta Outflow Requirements (maf/year)

more uniformly across all regions using waters tributary to the Delta.[9] Although urban water scarcity would increase as the regulations on Delta outflows became stricter, average scarcity levels would never exceed 100 taf per year (an amount too small to be seen in Figure 6.4). With stricter regulations, Sacramento Valley water might be sold in greater volumes to users south of the Delta.

As seen in Figure 6.5, small increases in minimum net Delta outflow would lead to fairly small cost increases as long as water resources were managed cost-effectively. However, as these requirements increase further, water scarcities would affect more highly valued crops and a few more urban water users. At the highest feasible levels of required minimum net Delta outflows (19.2 maf per year), water scarcity costs would approach those for ending water exports entirely ($900 million per year versus

---

[9]As before, Southern California agricultural users are unaffected, because their Colorado River supplies are isolated from Delta effects because of the limited capacity of the Colorado River Aqueduct.

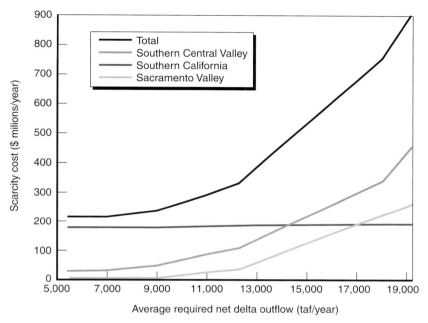

SOURCE: CALVIN model.

Figure 6.5—Average Annual Water Scarcity Cost by Region with Increasing
Minimum Monthly Net Delta Outflow Requirements

$1,041 million per year, respectively). However, annual statewide water deliveries would be much lower (29 maf versus 34 maf, respectively). This comparison illustrates the economic value of being able to share water deliveries across the state; moving water across the Delta substantially diminishes the economic effects of any reductions in total water deliveries.

The greater flexibility of the increased minimum outflow plan would make it less costly than the no-export alternative to maintain existing wetland wildlife refuges in the San Joaquin Valley. Both the no-export and the increased minimum outflow alternatives have the potential to offer additional benefits in terms of increasing terrestrial ecosystem habitat restoration on the western side of the San Joaquin Valley. Some reductions in irrigated land area might serve this purpose, provided that these lands are not too salinized from years of agricultural use.

Adapting Delta Agriculture to Salinity Changes

In-Delta agriculture depends on the availability of land and water supply. As seen in Chapter 2, the salinity of Delta water supplies has been a primary concern for Delta agricultural interests for at least a century. The DAP model can estimate changes in cropping patterns as well as farm revenues and profits that would occur under various management strategies that may increase the salinity of some parts of the Delta. Figure 6.6 shows the estimated distribution of farm revenues (per acre of agricultural land) for each Delta island for typical, current summer salinity conditions. Currently, the economic value of agricultural production is not uniform throughout the Delta, and agricultural production in much of the western and central parts of the Delta is quite low. Total agricultural revenues for this base case scenario—intended to simulate current conditions—are $367 million per year, with profits estimated at $201 million per year.

Figure 6.7 shows the economic value of agricultural production revenues for each Delta subregion when salinities are 10 times higher than in the base case conditions. This tenfold increase in Delta salinity would reduce overall agricultural revenues to $329 million per year, a decline of $38 million per year or roughly 10 percent. Profits would be reduced by almost 12 percent ($34 million per year) to $178 million per year and irrigated land area would be reduced by about 2 percent (less than 6,000 acres). The model suggests that these higher salinities would not end agriculture on any island. The agricultural economic effects of any Delta salinity scenario can be estimated in this way.[10]

Certainly, much higher salinity scenarios are possible. The DRMS examined a many-island levee failure that resulted in much higher salinities far into the Delta for one year. The DAP model may be adapted to estimate the agricultural economic effects of such emergency scenarios as well as

---

[10]See Appendix Figure D.4 for the corresponding results for a twentyfold salinity increase. Relative to the base case, a twentyfold increase in salinity reduces overall annual crop revenues to $254 million per year (–31%) and profits to $135 million (–33%).

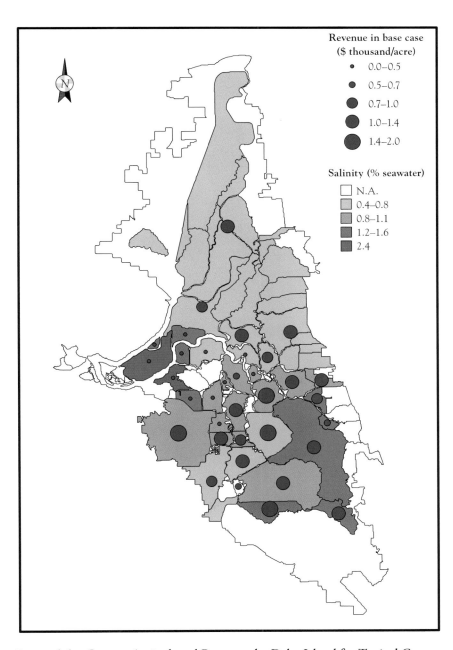

Figure 6.6—Current Agricultural Revenues by Delta Island for Typical Current Salinity Levels

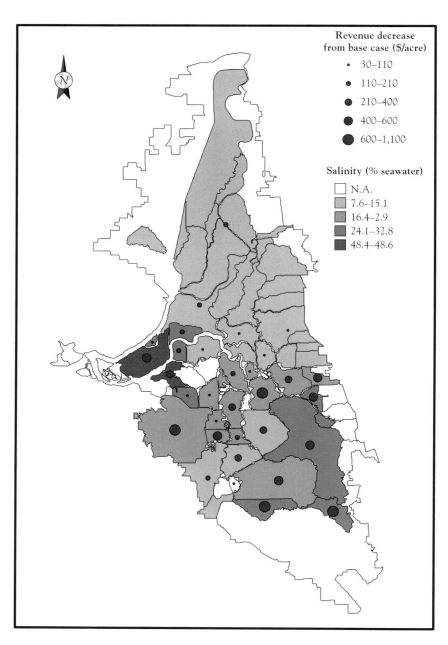

Figure 6.7—Agricultural Revenue Decreases with a Tenfold Increase in Delta
Salinity Levels

more typical salinity patterns that might be expected under scenarios with managed salinity fluctuation. A well-managed salinity fluctuation regime for the Delta should be able to avoid catastrophic scenarios.

Even today, salinity is not uniform throughout the Delta; substantial amounts of salt are introduced with the tides and with the agricultural drainage in San Joaquin River flows. If salinities in some parts of the Delta are allowed more variability to support desirable species, only some parts of the Delta are likely to be affected. In particular, western, central, and southern parts of the Delta would see the greatest effects. However, the large freshwater inflows likely from the Sacramento River would keep northern areas of the Delta rather fresh and unaffected by salinity from seawater or from San Joaquin Valley drainage under almost any conditions.

The ability of Delta farmers to support local levees is already severely limited by the profitability of this land use. The least profitable islands tend to be those in the western and central Delta, the most desirable areas for reintroducing salinity fluctuations. Reductions in profit from higher salinity in these areas would further reduce farmers' abilities to support levee improvements, requiring either additional state subsidies or eventual abandonment of these levees.

Modeling can be used to estimate the effects of changes in salinity on agricultural production and profitability within the Delta, and to help design mitigation strategies. As seen in Figure 6.7, the costs of a tenfold increase in salinity are not evenly distributed across Delta islands but are instead concentrated in areas of the Delta that already have higher salinities from tidal seawater mixing and San Joaquin River drainage. More detailed hydrodynamic modeling studies would be required to estimate salinity conditions specific to various water management alternatives. Models such as DAP can then be used to estimate the effects of management strategies for in-Delta water users in ways comparable to the economic and management evaluations modeled with CALVIN for areas outside the Delta.

Water Supply Aspects of a Peripheral Conveyance System

One of the most discussed "solutions" for the problems of export water supplies from the Delta is the so-called peripheral canal. As we saw in

Chapter 2, since the 1940s, various alternatives have been proposed to construct an isolated canal from the Sacramento River to south of the Delta, as a way to bypass the operational, water quality, and environmental problems associated with conveying exports through the Delta itself. During the 1960s and 1970s, one such proposal was promoted as a component of the State Water Project, but it was soundly rejected by voters in 1982. In light of concerns over the fragility of Delta levees, some exporters have recently revived the idea of a canal, and a Bay Area legislator has formally proposed such a facility (see Chapter 5). Benefits often cited in regard to a peripheral conveyance facility include:

Increased water export reliability. In earlier proposals for a peripheral canal, a key objective was greater operational flexibility, which would permit increased export quantities under many conditions. More recently, the peripheral canal proposal has resurfaced as a way of maintaining export capability without depending on fragile and seismically vulnerable levees or necessarily increasing export levels.

Improved export water quality. A peripheral conveyance facility would avoid contaminants that appear in Delta flows, which arise from in-Delta agriculture and urban activity, San Joaquin River drainage, and seawater. This objective is particularly important for urban water agencies, which face increasingly stringent requirements for drinking water treatment and regulation of disinfection by-products.

Reduced fish loss from Delta pumping. As early as the 1970s, some biologists saw such a peripheral canal as a way to reduce entrainment of pelagic fish and other organisms and to decrease confusion in the fish migrations that result from in-Delta pumping (Arnett, 1973). Recent work on pelagic organism decline indicates that Delta pumping may play a significant role in the decline of delta smelt (William Bennett and Wim Kimmerer, 2006, personal communication).

More natural in-Delta circulation and mixing. Recently, other ecological benefits of a peripheral conveyance system have been recognized. Such a system would allow water flow and quality in the Delta to vary more naturally. As discussed in Chapter 4, this change in circulation could be important for some native fish species in the Delta.

Overall, the primary benefit of a peripheral canal is the flexibility it would provide for combining water supply and ecological operations, which are currently antagonistic. Such a facility would break the connection between water exports and the maintenance of a homogeneous freshwater Delta. Greater operational flexibility would be available to manage diversified habitat in various parts of the Delta. With a peripheral canal, it would become easier to allow water flow and quality to vary in different parts of the Delta, perhaps increasing the overall suitability of the Delta for desirable species. With variable water conditions in the Delta, it is also possible to envisage continued use of occasional direct pumping from the Delta, for instance, during wet conditions.

There are many possible peripheral canal alternatives, with a wide range of details, including flow capacity, fish screening, inlet locations, outlet locations, routing, environmental mitigations, operation policies, ownership, and finance. Unfortunately, most analytical capability for water management in California is not currently suited to examining these alternatives, particularly if the goal is to manage variable conditions in the Delta. Current Delta hydrodynamics modeling capability is not suited to the study of significant changes from current Delta island configurations and conditions. The CALVIN model does not represent environmental and water quality aspects in enough detail to examine most peripheral canal alternatives.

However, analysis of adjustment costs under the management alternatives examined above does permit approximations of the value of a peripheral canal for water exporters. For the no-export case, the value of allowing a small amount of exports averages almost $1,300 per acre-foot, permitting reduction or elimination of expensive seawater desalination in the Bay Area. For cases in which environmental restrictions limit direct exports from the Delta, the value of a peripheral canal could be a few hundred dollars per acre-foot. Additional benefits would accrue in terms of export water quality. The DAP model results provide some indication that the costs to Delta agriculture need not be catastrophic, even if the canal resulted in some increases in Delta salinity levels. As seen above, a tenfold increase in irrigation season salinity throughout the Delta results in an estimated 10 to 11 percent decrease in crop revenues and profits within the Delta. A twentyfold increase in salinity reduces revenues and profits by about one-third (Appendix D).

Conclusions

Most of California's urban and agricultural water users depend on the Delta for much of their water supply. This broad dependency makes the health of the Delta a major common concern for almost all major water users. Nevertheless, water users and managers generally have substantial capacity to respond to changes in Delta management, including such extreme strategies as the elimination of the Delta as a water source. Model results suggest substantial ability to adapt if preparations, such as conveyance interties and coordinating agreements, are made. Comparisons of our model results with results from the DRMS indicate that abrupt unprepared changes, or hard landings, are much more expensive for water users than are well-prepared changes, or soft landings. Many agencies are already taking steps to reduce their vulnerability to short-term and long-term losses of Delta water supplies. Most water management decisions are made by local agencies and water users, and a productive role for the state is to facilitate the use of local decisions and resources for common state and local purposes. In the current era, local agencies and users often have greater flexibility and financial resources, and greater expertise about local management options, than state and federal agencies.

Maintaining a freshwater Delta in the face of accumulating permanent or semipermanent levee failures and sea level rise would likely require additional net Delta outflows. Delta farmers and urban agencies that draw water directly from the Delta (notably the South and Central Delta Water Agencies and the Contra Costa Water District, respectively) are likely to call for such outflows to preserve fresh water in the Delta. This chapter explored two extreme management changes to achieve this goal: elimination of all direct Delta exports and great increases in minimum Delta outflows. Although these alternatives result in high regional economic costs and inconvenience, the costs are not catastrophic relative to the state's overall economy. The costs of planned elimination of Delta exports are large, but not catastrophic, for urban water users in Southern California and the Bay Area. However, eliminating exports would greatly reduce agricultural activity in the western San Joaquin Valley and Tulare Basin, with likely catastrophic results for some agricultural communities in these regions. The costs of increasing net Delta outflows are much lower, but this alternative would require considerable re-operation of groundwater and surface water storage south of the Delta, with some reductions in agriculture in the San Joaquin Valley and Tulare Basin as well as sales of water (and reduced agricultural production) in the Sacramento Valley.

Under any of these scenarios, the loss of water supplies to agriculture south of the Delta would change the character of many rural agricultural communities in that region. Many farmers with senior water rights or contracts may do well financially by selling water, but other farmers and local workers and businesses are likely to do less well. In this case, mitigations and compensations (discussed in Chapter 9) seem appropriate to ease the transition.

Delta farmers were among the earliest major water diverters in California. Many of the changes suggested in Chapter 4 could increase water salinity for farmers in some parts of the Delta. But farms are businesses. The DAP model provides a way to estimate the effects of changes in salinity patterns, allowing benefits to be compared with costs and potential mitigation expenses to be estimated.

Although existing analytical capabilities for evaluating the operation and performance of peripheral canal alternatives are poor, some qualitative observations can be made. These are not based on CALVIN modeling but on observations and understanding of system behavior. Foremost is that many forms of a peripheral canal would break the connection between moving water to southern communities and maintaining the Delta as a homogeneous freshwater environment, thereby allowing for more dynamic and spatially varied management of the Delta. Models such as DAP could be useful in assessing the likely effects of various spatially varied management solutions for in-Delta agriculture or other uses. Initial results indicate that Delta agriculture would not be eliminated by some increase in salinity, although it would face significant additional costs.

This examination of the water supply consequences of some extreme alternatives for Delta water management provides a useful contribution to a broad discussion of alternatives for the Delta, to which we turn in the following chapter. These modeling efforts also illustrate the potential of modern mathematical models for evaluating and identifying promising solutions to large-scale problems such as those facing the Delta. Without the use of computerized models, the systematic exploration, development, and comparison of integrated solutions are severely handicapped.

7. Delta Options and Alternatives

"We must dare to think 'unthinkable' thoughts. We must learn to explore all the options and possibilities that confront us in a complex and rapidly changing world. We must learn to welcome and not to fear the voices of dissent. We must dare to think about 'unthinkable things' because when things become unthinkable, thinking stops and action becomes mindless."

J. William Fulbright, March 27, 1964

As we saw in Chapter 2, alternatives for managing the Delta have been widely discussed from technical, economic, regulatory, and political perspectives for over a century. Over time, management objectives have evolved. Following the initial focus on flood control for reclaimed Delta islands in the late 1880s, the primary goals of the large water projects built between the 1930s and 1970s were salinity control for in-Delta agriculture and water supply for farmers and urban areas to the south and west of the Delta. Environmental concerns, particularly for the health of key Delta fish species, moved to prominence in the 1970s, and by the early 1990s they led to the creation of the CALFED process.

Some of the earliest examinations of management alternatives were the most thoughtful and in-depth, driven by salinity intrusion problems that resulted from greater urban and agricultural use of the Delta itself and increased upstream diversions (Jackson and Paterson, 1977). These studies, mostly conducted in the 1920s and early 1930s, focused almost exclusively on two approaches to salinity management: physical seawater barriers and "hydraulic" barriers, which would regulate net Delta outflow from reservoir releases to keep the Delta fresh (Table 7.1). These earliest examinations consisted of multiple volumes of detailed and probing technical and economic studies (Young, 1929; Matthew, 1931a, 1931b), and they were accompanied by the kind of intense political and policy debates that still characterize Delta discussions. In the 1950s and 1960s, a much more diverse range of approaches was considered; however, the depth of their technical and economic examination was more limited (Jackson and Paterson, 1977). The same could be said of the CALFED investigations

Table 7.1

History of Major Delta Alternatives Studied

Year	Delta Alternatives
1848– 1930s	Private and Reclamation District Development Channelizing and leveeing islands with federal navigation improvements
1931	California Water Plan, 1930 Various downstream seawater barriers Hydraulic barrier—net Delta outflow of 3,000–5,000 cfs[a]
1955	Board of Consultants Six downstream seawater barrier plans Upstream barriers and control structures for through-Delta conveyance (Biemond Plan)
1960	California Department of Water Resources Seawater barrier at Chipps Island Four through-Delta conveyance and barrier plans, variants on the Biemond Plan
1963	California Department of Water Resources Seawater barrier at Chipps Island Peripheral canal (22,000 cfs capacity) Hydraulic barrier "Typical Alternative Delta Water Project"—a through-Delta alternative
1980s	California Department of Water Resources Various barrier and flood control programs for the Delta
1996	CALFED Bay-Delta Program (various alternatives considered) Extensive demand management New storage to improve Delta flow Dual Delta conveyance[b] Through-Delta conveyance Delta channel habitat and conveyance Extensive habitat restoration with storage Eastside foothills conveyance Chain of lakes conveyance Westside conveyance and river restoration Eastside conveyance

Table 7.1 (continued)

Year	Delta Alternatives
2000	CALFED Record of Decision (current policy, with reassessment of goals and objectives in 2007) *Through-Delta conveyance maintained, with levee strengthening, water use efficiency, habitat restoration, and water operations features*

SOURCES: Jackson and Paterson (1977); CALFED (1996, 2000b).
NOTE: Elements in italics were implemented.
[a]Analyses in the mid-1940s included consideration of a peripheral canal.
[b]CALFED's dual Delta conveyance included a peripheral canal (10,000 cfs capacity) and through-Delta pumping.

conducted in the mid-1990s, which broadened the scope of enquiry but looked at most alternatives in a relatively cursory manner (CALFED, 1996, 1997).

Most recently, the Delta has yet again become a topic of urgent policy discussion, for numerous reasons: unease over continued ecological declines, renewed awareness of vulnerabilities to earthquakes and flooding, and increased concern for the effects of Delta water quality on urban and agricultural users, as well as urbanization pressures, sea level rise, and regional climate change. The policy response has included various agency, legislative, and private efforts to examine Delta alternatives, including a flurry of conferences, hearings, workshops, media assessments, and many fine speeches that typically focus on various "obvious" solutions to the Delta's problems. To date, however, there has been no effort to list and systematically evaluate the range of alternative futures for the Delta.

In this chapter, we review the central issues that any Delta alternative must seek to address. We then present nine alternative solution strategies for the Delta, composed of a range of elements and options that address these central issues. Our aim is not to present an exhaustive list. For a system as large and complex as the Delta, examining "all possible alternatives" would be an infinite enterprise. Instead, our goal is to highlight a broad range of potential approaches, drawing from some of the most commonly suggested proposals, some classic alternatives from the past, and some relatively new approaches. Our focus is on strategies for better adapting the Delta to California's long-term needs and reducing

California's vulnerabilities to catastrophes in the Delta rather than on crisis responses to short-term catastrophes or small reductions in risk.

The Four Central Issues

Solutions for the Delta typically revolve around four central issues: Delta salinity, in-Delta land and water use, water supply exports, and the ecosystem. For each issue, various options are possible, either exclusively or in combination, within different locations in the Delta or for the Delta as a whole (Table 7.2). Any management alternative for the Delta should address all four of these issues.

Delta salinity has been a major concern for over 80 years, since the City of Antioch's 1920 lawsuit against Sacramento Valley irrigators (discussed in Chapter 2). Salinity affects the potability and taste of urban water supplies, the productivity of irrigated land, and the viability of aquatic ecosystems. For many decades, the focus of policymakers concerned about salinity revolved solely around keeping the Delta fresh, and the policy employed (a hydraulic barrier of net Delta outflow at the Delta's western edge) resulted in a sharp salinity change near Suisun Marsh. More recent thinking, reflected in Chapter 4, holds that having seasonal or even interannual variability in salinity in parts of the Delta may better mimic the Delta's natural conditions and help limit the extent of invasive species, which tend to prefer stable salinity or relatively constant freshwater flows.

Land use is another important issue in the Delta. Currently, most land in the Delta is agricultural, but there is substantial urban land and increasing economic pressure to urbanize more of the Delta, particularly near major transportation routes. Various infrastructure routes (e.g., ship channels, railroads, highways, pipelines, and power lines) traverse the Delta and must be either supported, altered, or rerouted—all at significant cost. A range of environmental uses already exist or could be created on Delta islands to support aquatic and terrestrial wildlife. The Delta also has increasing value for recreation, such as boating and fishing. Freshwater storage is another recent suggestion for Delta land use. This freshwater storage plan proposes investing in strengthening internal levees on some Delta islands subsided below sea level, allowing them to be filled with

Table 7.2

Delta Issues and Options

Salinity Conditions	Delta Land Uses	Water Supply Exports	Ecosystem Components
Fresh	Agricultural	Year-round Delta	Open-water habitat
Brackish	Urban	pumping	Riverine habitat
Fluctuating	Environmental	Seasonal Delta	Freshwater wetlands
	Recreational	pumping	Tidal brackish water
	Freshwater storage	Peripheral aqueduct	Seasonal floodplain
	Infrastructure support	Through-Delta facilities	Upland habitat
		No exports	

water, on a tidal or seasonal timescale, to help water projects pump fresh water from the Delta. All of these land uses have different implications for water use and the quality of water required in nearby channels, the volume and quality of drainage, and economic sustainability. Fortunately, the Delta is large and diverse enough to support a mix of land uses.

Water supply exports from the Delta are a major cause of controversy. With or without exports, the Delta would have many serious problems with flooding, land subsidence, degraded habitat, invasive species, and water quality. Any solution must address water supply exports, but there are many approaches to providing or avoiding this function for the Delta.

Likewise, any solution must address the Delta as a home for habitats that support a wide range of organisms, including many at-risk species. Broad habitat types important in the Delta include pelagic fish habitat, wildlife habitat, fresh open-water habitat, different forms of wetlands, and sustainable agricultural areas (see Table 4.2). Management options and decisions will determine the abundance of each habitat type. A key challenge will be managing the habitats to support desirable, mainly native, species and to keep populations of undesirable invasive species at a low level.

Finally, cultural values are also likely to have an important role for Delta management, for historical, recreational, local, and tribal interests.

Elements of Any Solution

Given the broad range of services demanded of the Delta, it is unlikely that any single action can resolve the Delta's problems. Instead, a portfolio of actions is likely to be required. Table 7.3 lists many potential elements of a comprehensive solution. Unfortunately, many proposed Delta "solutions" often advocate only one of these elements, with little discussion of how it would benefit or suffer from inclusion in a package of actions seeking to achieve a wider range of objectives.

Although current Delta management pursues a wide range of goals and includes many of these elements, the system's long-term sustainability is in doubt. Elements not currently pursued are controversial in one way or another, as they represent change—in water exports, land use, or associated economic activity.

Delta exports and inflows. Water supplies to users upstream or downstream of the Delta can be addressed by several options listed in Table 7.3, alone or in combination. Exports can occur via pumping through the Delta (the present method) or via a peripheral conveyance channel. Since the 1940s, regulation of outflows has been a way to keep the Delta fresh. As we saw in Chapter 6, this type of regulation can affect all users of Delta waters, including exporters, in-Delta users, and upstream diverters on the Sacramento and San Joaquin Rivers. It is also possible to imagine constructing more and better fish screens, or changing operations, or otherwise reducing harm to fish from exports, or stopping water exports from the Delta completely.

Internal flow modifications. The Delta's sheer size and hydraulic complexity provide many opportunities for internal flow modifications to achieve water supply and water quality goals. These include a wide variety of minor and major physical and operational changes. Only a few potential changes to internal Delta hydraulics have ever been explored in great depth. Currently, temporary barriers in the southern Delta are used to help maintain a barrier during the summer and fall months. The South Delta Improvement Plan envisions the use of operable flow barriers to improve flows and water quality (CALFED, 2000a, 2000b). It is likely that some new internal modifications would be desirable as part of almost any long-term solution for the Delta.

Table 7.3

Elements of Potential Delta Alternatives

Delta Water Exports and Inflows

1. Year-round pumping within the Delta[a]
2. Seasonal pumping
3. Peripheral aqueduct from the Sacramento River
4. Extended South Folsom Canal from the American River
5. Regulation of Delta inflows and outflows[a]
6. Screening for power plant cooling water (currently resulting in substantial fish entrainment)[a]
7. Fish screens at pump intakes (currently not in place everywhere)[a]

Internal Flow Modifications

1. Channel barriers
2. Temporary barriers[a]
3. New channels and flow capacities
4. Alteration of existing channels
5. Locks
6. Tide-gates (one-way)
7. Operable gates
8. Relocation of water intakes
9. Floodways (using existing farmland)
10. Levee and island barriers[a]

Reductions in Salt and Contaminant Loads

1. San Joaquin Valley drain to western Delta
2. Reduction of salt loads entering the San Joaquin River
3. Reduction of pesticides and other toxicant discharges[a]
4. Reduction or modification of Delta island drainage

Levees

1. Current levees[a]
2. Upgraded current levees to PL 84-99 standards (CALFED goal)
3. Fortified levees
4. Setback levees (located some distance from shore, difficult for subsided islands)
5. Environmental levees (designed to improve ecosystem habitats)
6. Storage levees (levees with internal and structural modifications to enable water storage)

Delta Island Uses

1. Urban uses[a]
2. Agriculture[a]
3. Environmental uses[a]
4. Recreation[a]
5. Freshwater storage
6. Flood bypasses

Table 7.3 (continued)

Civil Infrastructure

1. Stockton ship channel[a]
2. Sacramento ship channel[a]
3. Railroads[a]
4. Highways, roads, and bridges[a]
5. Gas and water pipelines[a]
6. Electric power lines[a]
7. Underground gas storage tanks[a]

Mitigations

1. In-kind exchanges of water supplies or land
2. Financial compensations
3. Other types of transitional support

NOTE: These actions are representative; additional elements are possible.
[a]Currently in use.

Reductions in contaminant loads. Water quality in the Delta is severely compromised by the salts, pesticides, and nutrients that drain from San Joaquin Valley farms into the San Joaquin River; agricultural drainage from Delta islands adds to this problem. Urban runoff is also a contributing factor. Several approaches exist for addressing this problem. These include reductions in drainage flows, reductions in the salinity of water used for irrigation, greater dilution of drainage waters with cleaner water, and the construction of a drain to dispose of drainage water downstream of the Delta (similar to the Kesterson Drain concept).[1] Although some recent programs have begun to encourage farmers to diminish harmful runoff (for instance, through changes in pesticide use), the contaminant problem remains largely unresolved. Given the growing evidence that contaminants are harming Delta wildlife, it is likely that better pollution control will need to be part of any future Delta alternative.

---

[1]The San Luis Drain was built to convey drainage from westside San Joaquin Valley farms to the Kesterson Reservoir. It opened in 1981 but was closed in 1985 because the selenium (a highly toxic type of salt) was severely damaging wildlife in the area of the drainage ponds. A reformulated project, involving prefiltration of the toxic waters, is among the options being considered by the U.S. Bureau of Reclamation, which is apparently under legal obligation to provide a drainage solution for some CVP contractors (Boxall, 2006).

Levees. The backbone of the current system is 1,100 miles of Delta levees. Improvement of levee reliability and environmental performance may take many forms. Modification to some of the Delta's levees is likely to be desirable. It is probably not desirable to treat all levees in the same fashion.

Delta island uses. Land use decisions or regulations for a variety of land uses will be an indispensable part of any Delta solution. Different land uses create different requirements for flood protection, water quality, and transportation and have different implications for management costs, land subsidence, water use, drainage water quality, environmental performance, and sustainability.

Civil infrastructure. As noted above, the Delta's lands and waterways are also used as conduits for a variety of civil infrastructure. The navigation depth and channel geometries of the Sacramento and San Joaquin ship channels have important implications for hydrodynamics and water quality. The viability of specific configurations for roads, rail lines, bridges, and power and water pipelines depends on decisions about Delta levees and channels.

Mitigations. A long-term Delta solution would have to include some form of compensation for interests who cannot be reasonably satisfied in terms of their water or land use rights. As discussed further in Chapter 9, mitigation measures to ease transitions might include in-kind compensation, financial compensation, or other measures.

Although Table 7.3 does not provide an exhaustive list, it represents the type and range of activities that might be included in a more successful approach to managing the Sacramento–San Joaquin Delta. In the remainder of this chapter, we draw from this list to outline nine possible alternatives. Of course, many combinations of the elements on this list could result in viable alternatives, and it is impossible to examine all of them.[2] We chose these nine to illustrate some basic types of approaches, with the hope of improving the public discussion of Delta solutions and policies. Often, interested parties will seek to immediately identify what

---

[2]Even simple combinations of only 20 elements in Table 7.3 result in 2^{20} = 1,048,576 alternatives.

they see as the "obvious" solution to the Delta's problems. At this time, we do not see any strong comparative basis for such assertions.

Delta Alternatives: A First Cut

Our nine potential Delta alternatives consist of some prominent contemporary solutions, some long-standing historical solutions, and some novel recent solutions. The solutions fall into three broad categories (Table 7.4). The first category includes alternatives that aim to maintain the Delta as a freshwater body, consistent with policies pursued over the past 70 years. The second category includes alternatives that continue to allow substantial water exports, but with some basic changes in water management to allow for fluctuating salinity and local specialization of Delta land and water uses. A third set of alternatives considers changes that substantially reduce or modify the role of exports. Although not exhaustive, these nine alternatives should suffice to illustrate the breadth of solutions that might be available. We discuss the broad contours of each alternative below. In Chapter 8, we provide a preliminary comparative evaluation. However, before any long-term decisions are made, more detailed specification, design, and evaluation are needed.

Freshwater Delta Alternatives

In these alternatives, the Delta would be maintained as a largely freshwater body, and all water exports would continue to be made directly from the Delta. For decades, water managers and interests have sought solutions to maintain these objectives, including the constructing of physical salinity barriers and hydraulic barriers of various forms. Although maintaining the Delta as a freshwater body provides considerable water supply convenience for water users in and south of the Delta, it implies reliance on levee structures or salinity barriers as well as upstream reservoirs with sufficient inflows to restrict seawater intrusion. A levee-dominated solution does not automatically imply a freshwater Delta, but maintenance of Delta levees has become associated with supporting fresh water use for exporters and in-Delta pumpers.

1. **Levees as Usual.** This is a business-as-usual Delta. The current levee-intensive system would be maintained with something close to

Table 7.4
Nine Long-Term Delta Alternatives

Alternative	Delta Salinity	Delta Land Uses	Water Supply Exports
Freshwater Delta			
1. Levees as Usual—current or increased effort	Fresh	Agriculture	Delta pumping
2. Fortress Delta (Dutch model)		Agriculture and urban	
3. Seaward Saltwater Barrier		Agriculture and urban	
Fluctuating Delta			
4. Peripheral Canal Plus	Fluctuating (west) and fresh (east/south)	Mixed, locally specialized	Peripheral aqueduct
5. South Delta Restoration Aqueduct		Mixed, locally specialized	Peripheral aqueduct
6. Armored-Island Aqueduct		Mixed, locally specialized	Delta pumping
Reduced-Exports Delta			
7. Opportunistic Delta	Fluctuating and fresh	Mixed, locally specialized	Delta pumping (variable)
8. Eco-Delta		Mixed, locally specialized	Delta pumping (reduced)
9. Abandoned Delta		Trend to open water	No exports

NOTES: "Mixed" includes agricultural, urban, and environmental land uses. Other alternatives and combinations could be conceived of.

recent levels of effort or modestly upgraded to meet the federal PL 84-99 standards for agricultural levees (CALFED, 2000a). Failed levees would be repaired to prior conditions, along with most flooded Delta islands. Delta management would be crisis management—dealing with system failures and deterioration—but increased investment in levees would reduce crisis frequency. This approach could become expensive; for example, in 2003, state financial liabilities for levee failures increased as the result of the *Paterno* decision, which made the state liable for flood damage behind "project" levees belonging to the Central Valley flood control system (Department of Water Resources, 2005a). Levee failures may occur individually, for no particular reason, or in groups as a result of floods or earthquakes. Although this alternative continues to provide an inexpensive short-term solution for some users of Delta services, any levee failures will result in either increasingly expensive levee maintenance and island reconstruction costs or increased numbers of flooded and abandoned islands (such as today's Franks Tract and Mildred Island). As levee failures accumulate, the Delta ultimately becomes a collection of flooded islands.

2. **Fortress Delta.** In this alternative, "whatever it takes" investments would be made for constructing, maintaining, and repairing levees, investing in considerably more than the 200-year level of protection for urban and urbanizing areas of the Delta (which can afford such protection) as well as in high levels of protection for selected Delta islands critical to maintaining a freshwater Delta. These levees would be upgraded and maintained on the Dutch model, where design floods range from the 1,250-year to 10,000-year events (Van Der Most and Wehrung, 2005).[3] To make this effort more cost-effective and reliable, the total length of levees in the system would be shortened, reconfiguring some islands. Fortification efforts would especially focus on western islands and would include seismic upgrades to both embankment materials and levee foundations. Many interior islands would not be fortified, unless deemed necessary for protecting urban areas or for providing barriers for salinity encroachment into the Delta.

---

[3]Note that Dutch and American calculation methods differ for estimating flood frequency. Infrequent floods typically appear more infrequent when using Dutch estimation methods (personal communication, Joe Countryman, MBK Engineering, 2006).

Over time, the lower-reliability levees in the Delta's interior would be likely to fail. Failed levees on many central and eastern islands would probably not be repaired, given the costs relative to the value of their previous land uses and the lack of need to maintain them for water export quality. This would provide for a gradual, if somewhat random, increase in open-water habitat over time. Figure 7.1 illustrates one such alternative.

3. **Seaward Saltwater Barrier.** Seaward saltwater barriers are one of the oldest and most extreme proposals for maintaining the Delta as a freshwater body (Young, 1929; Matthew, 1931a, 1931b; Jackson and Paterson, 1977). This type of solution was endorsed by many agencies in the past, mostly before 1963 (see Chapter 2). Most seaward salinity barrier proposals have recommended building locks for ship passage and gates or spillways for passing major floods, with the major goal of providing reliable freshwater quality upstream of the barrier. A complete seawater barrier would also turn the Delta into a freshwater reservoir. With the current configuration of islands, the usable storage capacity would likely be small (about 100,000 acre-feet), although reservoir capacity could increase as levees fail. Investigations by the Army Corps of Engineers' Waterways Experiment Station in the late 1970s considered partial barriers, such as underwater sills in Carquinez Strait, to restrict seawater flows into the Delta. Over the past year, several Dutch engineers have suggested the construction of a large movable barrier similar to the Maeslant storm surge barrier that protects Rotterdam in the Netherlands (Breitler, 2006).

In the past, problems with expense, navigation, Delta island levee failure, water quality, and fish passage led to the rejection of seaward saltwater barriers. Such impediments are likely to be even greater today, given heightened concerns about fish passage, connectivity among habitat areas, and polluted urban and agricultural runoff. However, on a smaller scale, salinity barriers may have some potential for regulating tidal flows and salinity in parts of the Delta. For instance, a small saltwater control structure was constructed on Montezuma Slough in Suisun Marsh in 1988. Temporary barriers also have become common in southern parts of the Delta.

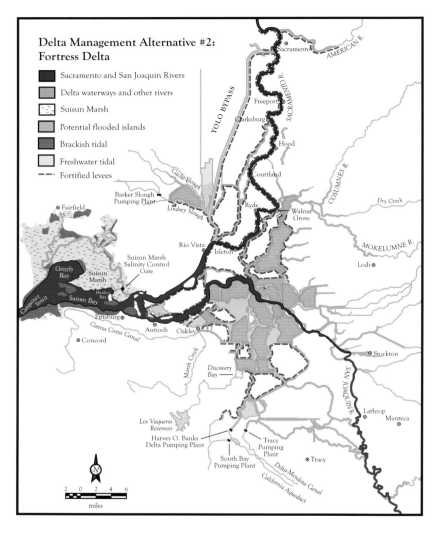

Figure 7.1—Delta Management Alternative #2: Fortress Delta

Fluctuating Delta Alternatives

By hardening water export capacity within the Delta itself or through a peripheral canal, parts of the Delta could feature fluctuating salinity to promote desirable species, while other parts remain fresh. Such alternatives would allow local areas within the Delta to take on more specialized ecosystem and economic functions, and certain current functions could change location. For example, duck clubs in the Suisun Marsh area would shift to western and central Delta islands, allowing Suisun Marsh to specialize in fish and wildlife that require more naturally fluctuating salinity conditions. The Yolo Bypass and Cache Slough area would be managed for greater fish-rearing habitat. Eastern and southern peripheral islands and lands with better transportation access would have more urban development, which would finance Dutch-standard urban protection levees. Many Delta islands would remain agricultural, with most Delta recreation remaining intact, although, again, there would be some rearrangement. Some of the more subsided islands might be flooded or allowed to flood (with or without levees) for water storage, fish habitat, or both. Some levees might be breached in a planned manner, whereas others might be allowed to fail, allowing continued near-term agricultural production, avoiding long-term state financial liabilities, and providing a long-term means of land use change. Salinity in the western Delta would become more naturally fluctuating but would remain fresh for much of the year. The eastern Delta would remain fresh except for salt loads from San Joaquin Valley agriculture.

To allow salinity to fluctuate within the Delta for ecosystem purposes, other provisions would be made for Delta water exports. Here, we consider two variants of a peripheral canal and one through-Delta conveyance facility. In light of the frequent discussions of the advantages and disadvantages of a peripheral canal, it is worth noting some general considerations at the outset.[4] For water exporters, a peripheral canal represents greater assurance of water quality (particularly lower salinity and disinfection by-product precursors) and somewhat greater assurance of quantities of water deliveries (because exports would be less susceptible to conditions within the Delta). Environmental groups often express some interest in a peripheral canal because, if properly operated, it should result in less disruption in fish migrations and should entrap or entrain far fewer fish (presuming the construction of adequate fish screens).[5] Such

---

[4]See also the discussion in Chapter 6.

[5]One option is to use the river bank as a filter or fish screen, a method known as "bank filtration." It is usually developed by placing a big well and pump near a porous

an upstream diversion for major Delta exports would also provide greater flexibility for regulating local in-Delta water flow and quality conditions. The two peripheral canal variants presented here illustrate some of these and other benefits, and point to certain precautions that could be taken in their construction and operation.

4. **Peripheral Canal Plus.** This alternative builds on the now-traditional concept of constructing an isolated facility or peripheral canal from the vicinity of Hood, on the Sacramento River, to the CVP and SWP canal intakes at or near Clifton Court Forebay. The canal would be supplemented by actions to improve conditions within the Delta for various purposes (ecosystem, recreation, agriculture, housing, etc.). The original peripheral canal proposal was for roughly 22,000 cfs (Figure 7.2 illustrates this proposal, which went to voters in 1982). A future canal might be much smaller, perhaps only supplementing continued direct exports from the Delta. The canal examined by CALFED (1999) considered a capacity of only 10,000 cfs. A larger canal would provide economies of scale and increase operational flexibility but would be limited by the combined downstream capacities of existing CVP and SWP export aqueducts (about 15,000 cfs). Operational flexibility includes the ability to manage salinity for ecosystem support. However, even a smaller canal might raise fears and concerns for water quality within the Delta. As large reductions in direct Delta export pumping would likely leave southern Delta channels stagnant, a mitigation or flow augmentation program might be needed to maintain water quality at a level required by Delta fish species, farming, and recreation. The precise package of noncanal activities would vary with desired in-Delta objectives.

5. **South Delta Restoration Aqueduct (SDRA).** The SDRA would consist of a canal similar to the one discussed in the previous example, but its major outlet would enter the lower San Joaquin River, perhaps as far upstream as Old River. Figure 7.3 illustrates one possible configuration of the SDRA. This canal would shift a portion of the Delta inflows from the Sacramento River to the San Joaquin River. These

---

river bank. Sometimes a big ditch or "infiltration gallery" is constructed parallel to the river to improve efficiency.

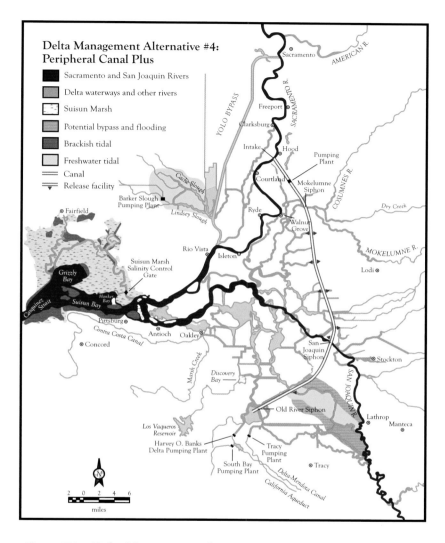

Figure 7.2—Delta Management Alternative #4: Peripheral Canal Plus

supplemental freshwater flows would resolve various water quality and flow problems of the lower San Joaquin River, the Stockton ship channel (which has seasonally low dissolved oxygen), and the southern Delta, while providing fresher water for ultimate export pumping. If these flows were introduced far enough up the San Joaquin River and

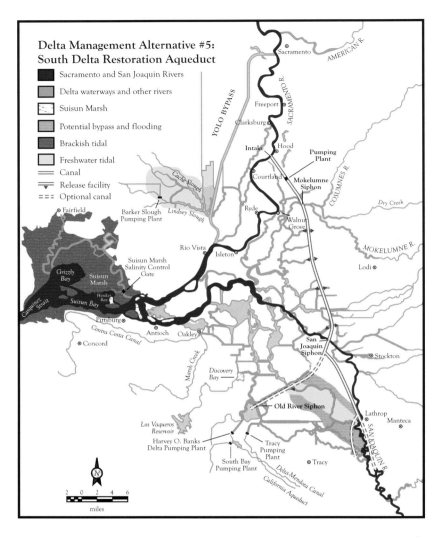

Figure 7.3—Delta Management Alternative #5: South Delta Restoration Aqueduct

additional channel changes were made, some of these flows could bypass the Stockton ship channel and go into a wetland and flood bypass channel through the southern Delta, contributing to improved habitat and agricultural water quality in that region. This alternative would also relieve some pressure on the Stanislaus River and other tributary reservoirs to achieve San Joaquin River water quality standards and environmental flows employed to help young salmon migrate down the San Joaquin River and through the Delta. Flows up to 5,000 cfs might be needed for the lower San Joaquin River to meet the objectives of the SDRA. Since this canal—unlike the peripheral canal alternative noted above—would rely on existing Delta pumping plant intakes for exports to points south, it would be subject to similar regulatory controls and restrictions. One variant of this alternative would be to have a smaller branch of the aqueduct directly feed high-quality water into the California Aqueduct and the Contra Costa Canal for urban uses.

6. **Armored-Island Aqueduct.** This is a through-Delta alternative in which a major semi-isolated freshwater conveyance corridor would be created by armoring selected islands and cutting off or tide-gating various channels within the central-eastern Delta. The location of this aqueduct would be determined on the basis of cost; seismic risk; water quality; Delta land use; and ship, boat, and fish passage considerations. (For an illustration, see Figure 7.4.) An armored-island aqueduct would allow restoration and reconfiguration of western islands and urban development on higher-elevation eastern lands and islands. Water exports might be supplemented with a through-Delta canal at Snodgrass Slough or a northeast Delta floodway at Tyler or Staten Islands. Intakes at the upstream end would need to be screened to prevent fish entrainment. It would be potentially problematic or expensive to maintain adequate depth where the aqueduct crosses the Stockton Ship Channel. Several forms of this solution were considered in the 1950s and 1960s as variants of the Biemond Plan (Jackson and Paterson, 1977), in the 1980s as the Orlob Plan (Orlob, 1982), and in the 1990s by CALFED (1996) as various through-Delta alternatives.

Reduced-Exports Alternatives

Several Delta alternatives rely neither on new Delta export facilities nor on levees. However, they imply an ability to greatly modify the pattern and quality of Delta exports. Two of the alternatives examined below would create a locally specialized Delta with fluctuating salinity, as in the

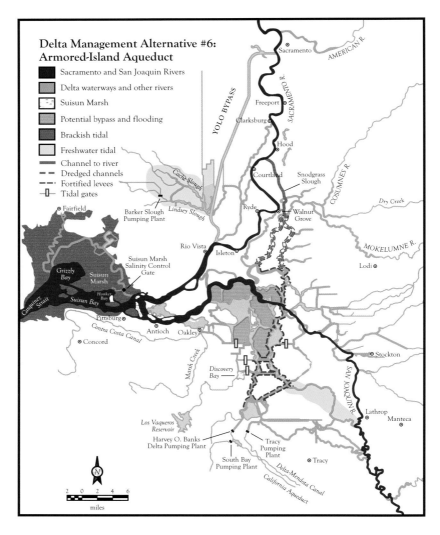

Figure 7.4—Delta Management Alternative #6: Armored-Island Aqueduct

preceding group. A third alternative consists of abandoning the Delta for most human purposes.

7. **Opportunistic Delta.** This alternative would allow opportunistic seasonal Delta exports only, during times of high discharge of fresh water in the Delta (generally the winter and spring months). This

change in pumping regimes might be accompanied by an expansion of export pumping capacities to allow larger volumes of water to be captured during the wet season. Such operations would allow greater natural fluctuations in western Delta salinities, which may have significant ecological value. Surface storage within and near the Delta might be desirable for this situation, allowing large gulps of fresh water to be taken when available, to be released more slowly into the canals south of the Delta, which have limited capacity. Additional storage south of the Delta, probably in groundwater banks, might also be useful to cover dry years when little opportunistic Delta pumping is available. Major in-Delta levee expenses would not be needed for water exports. Instead, expenses would be required for strategically located storage and other water supply alternatives, such as wastewater recycling. Because many, if not most, islands would become flooded as a result of subsidence and levee deterioration, opportunities would exist to create habitat favorable to desirable fish species, especially on western islands (e.g., Sherman and Twitchell Islands). Figure 7.5 illustrates an Opportunistic Delta alternative.

8. **Eco-Delta.** Restoring the Delta to something resembling its historical conditions is not possible because of the irreversible nature of many past alterations, such as invasions of alien species and land subsidence. Future changes, resulting from sea level rise and regional climate change, also mean that the Delta will never again be as it once was (or is now). However, it may be managed to favor key Delta species— especially at-risk native fish and birds and species important for fishing and hunting—and other desirable ecosystem attributes. In this scenario, water extraction, transportation corridors, and other functions would be maintained as long as they did not seriously interfere with rehabilitation goals. Some water exports would occur, but probably less than in the Opportunistic Delta alternative.

Some components of this vision include (1) flooded islands that provide habitat for pelagic species such as the delta smelt and that discourage undesirable alien species, (2) inland islands managed as freshwater wetlands for duck hunting and other purposes, (3) islands managed for upland foraging habitat for sandhill cranes and other

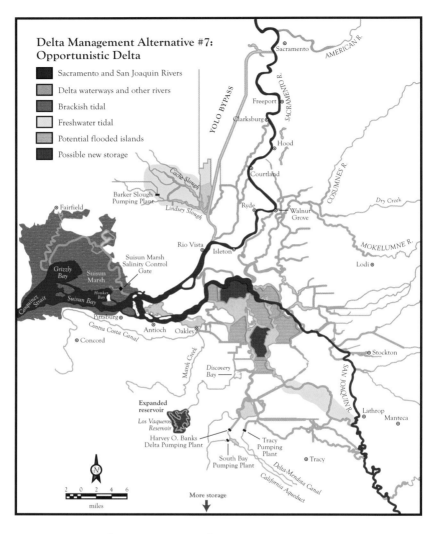

Figure 7.5—Delta Management Alternative #7: Opportunistic Delta

wintering waterfowl (presumably by wildlife-friendly farming), and
(4) large expanses of peripheral areas restored to some resemblance
of the historical Delta (e.g., Suisun Marsh, Cache Slough region,
Cosumnes River floodplain), as discussed in Chapter 4. The text box
below describes one possible configuration of Delta islands that would
manage the Delta mainly for ecosystem values; such a configuration

would also be consistent with several of the locally specialized alternatives discussed above. Figure 7.6 illustrates this configuration.

The Eco-Delta may also satisfy other goals. Strategic filling of subsided Delta islands is often suggested to enhance ecosystem restoration and levee stability. Island-filling opportunities might include restored tule marshes, seasonal or tidal freshwater storage to enhance water supply, carbon sequestration farms or parks to mitigate greenhouse gas emissions associated with economic activity, and disposal of dredged and other materials.[6] The Eco-Delta alternative would require a new administrative and financial framework for the Delta, along with significant changes in land use and ownership.

Current management of the Delta is not promising. However, because each Delta island can be put to different uses (or combination of uses), a nearly unlimited number of future alternatives exist. This text box illustrates one possible configuration of Delta islands that would manage the Delta mainly for ecosystem values (the Eco-Delta), in association with a peripheral canal (Peripheral Canal Plus or South Delta Restoration Aqueduct), or in association with opportunistic water withdrawals (Opportunistic Delta). The configuration draws on ecosystem needs in the Delta presented in Chapter 4. For an illustration, see the text box below.

9. **Abandoned Delta.** If the Delta proves itself to be an excessively unreliable or expensive part of California's water supply system, water users who currently depend on it can be expected to minimize or eliminate this dependency. Many Delta exporters already have taken steps to limit their reliance on Delta exports, with the development of conjunctive use and off-stream storage projects at Los Vaqueros, in the Tulare Basin, and in Southern California. Other activities under way or planned include local water demand reduction, water reuse, and desalination. In addition, Delta farmers, reflecting on the long-term capacity of the levees and increasingly saline irrigation water, may also plan to retire or move. Fishery agencies and interests, faced with the

---

[6]Carbon sequestration would work much as the pre-European peat swamp, taking carbon dioxide from the atmosphere into marsh plants or perhaps fast-growing trees. These plants could then be interred.

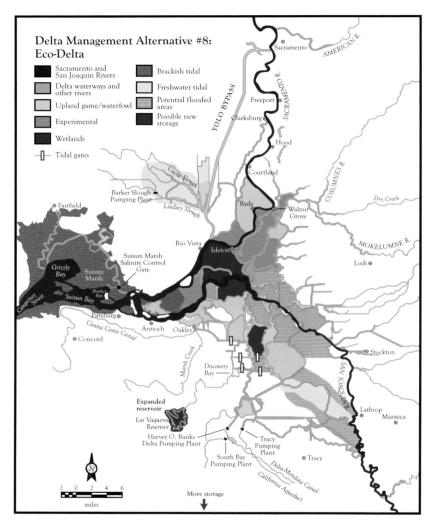

Figure 7.6—Delta Management Alternative #8: Eco-Delta

unreliability and seeming ineffectiveness of in-Delta restoration efforts, might seek to invest their limited resources elsewhere.

A planned multidecade retreat from the Delta might involve the eventual conversion of the western Delta and Suisun Bay to large patches of open water with fluctuating salinity, the transition of water suppliers to different supplies and additional water use efficiencies, and the phasing out of much of the Delta's farm economy. A slow unplanned retreat from the Delta, involving the cumulative effects of individual water user and landowner actions, is likely to provide a much less predictable outcome.

Heterogeneous Island Management

Current management of the Delta is not promising. However, because each Delta island can be put to different uses (or combination of uses), a nearly unlimited number of future alternatives exist. This text box illustrates one possible configuration of Delta islands that would manage the Delta mainly for ecosystem values (the Eco-Delta), in association with a peripheral canal (Peripheral Canal Plus or South Delta Restoration Aqueduct), or in association with opportunistic water withdrawals (Opportunistic Delta). The configuration draws on ecosystem needs in the Delta presented in Chapter 4. For an illustration, see Figure 7.6

1. Van Sickle Island would be flooded, as part of a general conversion of Suisun Marsh to a brackish tidal system.
2. Sherman Island would be managed as a patchwork of plots with various management objectives and experiments but would basically maintain its present configuration of levees.
3. Twitchell and Brannan-Andrus Islands would become islands in the style of the Delta Wetlands proposal, with the capacity to control flows in and out. A ring levee would surround the town of Isleton.
4. A levee would be constructed across the low-lying portion of Staten and Grand Islands so that the upper portions could be managed for sandhill cranes and for supporting agricultural practices that reduce land subsidence (e.g., rice farming).
5. Islands more than 15 feet below sea level—Bradford, Webbs, Bouldin, Venice, Empire, Rindge, McDonald, Medford, Mandeville, Bacon, Woodward, Lower Roberts—would largely be "let go" to become open-water habitat similar to Franks Tract.
6. Jersey Island, nonurban parts of Bethel Island, and Jones and Holland Tracts would be managed as waterfowl/wildlife islands, with Delta Wetland–style levees.
7. Hastings Tract and other lands in the Lindsey-Cache Slough regions would be managed as tidal freshwater (occasionally brackish) habitats.
8. Upper Roberts Tract and Union Island would be managed as tidal marsh habitat and as flood bypasses.
9. Other islands would be maintained under present uses, mainly agriculture.

The sudden abandonment of the Delta for water exports could also occur from the failure of many Delta levees because of floods or earthquakes, with grave consequences for those relying on the Delta's services. An abandoned Delta would likely have additional water quality problems in its southern and perhaps eastern areas. Aside from effects on landowners within the Delta, almost all adaptation expenses would be incurred outside the Delta. The eastern Delta ecosystem would most likely resemble that found on Franks Tract and Mildred Island, dominated by invasive, alien species such as Asiatic clams and Brazilian waterweed. The salinity regime in the western Delta would revert to greater fluctuation than at present. Nonwater-related uses of the Delta, for roadways and bridges, pipelines, and power lines, would be rerouted or hardened for these new conditions.

Some Unexamined Alternatives

We have discussed only nine of a nearly infinite number of possible Delta alternatives. Examining all possible alternatives is obviously impossible in the pure sense. Instead, our intent is to stimulate comparative, solution-oriented discussions of future options.

Hybrid Solutions

Many of the alternatives described above have promising features that could be combined into an even more favorable hybrid alternative. A more extensive study of solutions for the Delta should include the development, evaluation, and discussion of such alternatives. We suspect that the best solutions will be combinations of the ones described here, providing better performance across multiple dimensions, conditions, and a broader financial and political base. Identification of such alternatives is unlikely to emerge from a political process, however. A serious technical process, supporting a political process at some distance, will be needed.

Upstream Storage

We have deliberately avoided considering one set of commonly discussed alternatives that focus on the construction of additional upstream storage. As discussed in Chapter 2, the release of freshwater flows from upstream storage (particularly Shasta, Oroville, Folsom, and New Melones Reservoirs) has been a central tool in the regulation of Delta salinity since the early conception of both the Central Valley Project and the State Water Project. A persistent popular and political school of thought continues to support this strategy, in the belief that additional upstream storage capacity should remain part of the solution to problems with Delta salinity. However, there is little technical or economic support for expanding upstream storage to serve California's larger water system.

As discussed in Chapter 6, major expansions of upstream storage have scant likelihood of being an economically desirable solution, either on their own or as a central component of a Delta alternative. Because the biggest long-term problems within the Delta are island subsidence and levee weakening, the regulation of upstream flows would, by itself, be ineffective in resolving Delta salinity and flood control problems. The desirability of greater fluctuations in western Delta salinities further decreases the value of upstream reservoir storage. Although some expansions of storage capacity might have significant operational or water quality benefits for downstream water users, this is more likely to be appropriate as off-stream storage in locations south of the Delta. Even in these cases, off-stream storage probably would be cost-effective only for urban water users. We are unaware of *any* major recent study indicating that major reservoir expansion is economically justifiable in California for water supply purposes relative to other, more readily available forms of water supply. The fact that water agencies have not expressed a willingness to pay for storage projects, as they did for the development of the State Water Project, is another indication of the limited value of storage expansion relative to other investment opportunities. Serious discussions and policy debates on Delta water policy can ill afford to be distracted by efforts to include expensive and ineffective options as a major part of the solution strategy.

What's "New"?

There are few completely "new" Delta options. Over the decades, although many people have claimed to have found the "obvious" solution to the Delta's problems, disagreement tends to arise over which solution is "obviously" the best. Nevertheless, several relatively new ideas appear among the alternatives presented above.

- **Creating localized specialization within the Delta.** Traditionally, policymakers have sought to treat the entire Delta homogeneously. By maintaining the entire Delta as a perennially freshwater body, many habitats that once existed in the Delta have been displaced, and variability, which is useful for reducing the potential harm of invasive species and providing habitat for native species, has been reduced. Allowing different parts of the Delta to specialize in particular functions or services might allow for greater overall performance for all, or almost all, purposes. Local and temporal variability in flows and various aspects of water quality and habitat was common in the pre-European Delta. As discussed in Chapter 4, different areas of the Delta could specialize in supporting different types of habitat, with greater and more natural fluctuations in flows and salinity. One version of a heterogeneous Delta appears in Figure 7.6; it might apply to several of the Delta alternatives.
- **Establishing a western Delta fluctuating salinity ecosystem.** Western Delta salinity appears to have naturally fluctuated more in the past than it does now; reintroducing this fluctuation in parts of the western Delta should benefit many desirable species. Many of the alternatives proposed above would allow for greater fluctuations in salinity.
- **Using peripheral areas to bring back desirable natural conditions that existed in the Delta historically.** Suisun Marsh, Cache Slough, and Yolo Bypass are especially promising examples of locations that could serve valuable, but very different, environmental functions. Again, many of the above alternatives would allow for the return of natural conditions to parts of the Delta.
- **Allowing the urbanization of some Delta lands.** Local land use pressures, access to major transportation and employment centers, and financial opportunities make urbanization of some Delta lands seemingly inevitable, despite high costs and risks of flooding. Given recent housing prices, urbanization provides a significant ability to contribute financially and politically to solving problems in certain areas of the Delta and to aid the overall health of the Delta. Careful regulation should be able to provide substantial flood protection

(exceeding 200-year average recurrence) and prevent unreasonable interference with environmental functions.

- **Building a Sacramento–San Joaquin Canal.** Such a canal, a central feature of the South Delta Restoration Aqueduct alternative described above, would supplement lower San Joaquin River flows with Sacramento River water. This would provide larger flows than the San Joaquin supplemental flows envisioned in earlier peripheral canal proposals, because most or all canal flows would transfer into the San Joaquin River. Having Sacramento River flows enter the lower San Joaquin River should reduce the need for San Joaquin and Stanislaus River flows to improve water quality in the southern Delta and lower San Joaquin River.

- **Creating a San Joaquin River marsh and flood bypass.** As part of supplementing water deliveries to the San Joaquin River, a marsh and flood bypass system might provide additional environmental habitat, water quality improvements for southern Delta farmers, flood control capacity for the lower San Joaquin River, and conjunctive management opportunities (groundwater banking).

- **Managing expectations and providing mitigation solutions.** It is unlikely that any alternative would satisfy all Delta interests in terms of water and land use. The approach outlined here differs from the underlying assumption of CALFED that everyone can "get better together." Stakeholders whose land and water interests cannot be directly satisfied may be compensated by financial or other means. Even with such mitigations, one cannot reasonably expect universal satisfaction.

Conclusions

A primary thesis of this report is that variability in Delta flows, water quality, and functions is potentially desirable, allowing different parts of the Delta to function differently, as they did before European settlement. By insisting that all of the Delta be managed as a static system, as it is in its present configuration, a very unnatural Delta has been created—one that suits neither natural nor human objectives. Maintaining such a vast area, subject to great natural variability, as a more or less homogeneous region, requires substantial resources and implies substantial risks. The Delta is now too important to tolerate such risks.

The potentially catastrophic nature of risks in the Delta implies some need to seek solutions that allow for a soft landing. Any proposed solution will take considerable time to complete, but the existence of an agreed-on direction will allow California to take advantage of some opportunities and gradually transform the Delta into a more functional and less risky environment. In the next chapter, we evaluate the alternatives presented in this chapter. These evaluations are neither final nor highly detailed but are qualitative and based on information that is readily available at the present time.

8. Evaluating Delta Alternatives

"The true rule, in determining to embrace, or reject any thing, is not whether it have any evil in it; but whether it have more of evil, than of good. There are few things wholly evil, or wholly good. Almost every thing, especially of governmental policy, is an inseparable compound of the two; so that our best judgment of the preponderance between them is continually demanded."

Abraham Lincoln

As we saw in Chapter 2, early studies of the Delta sought solutions to meet a relatively narrow set of objectives: improving freshwater supply and reliability for water users within and south of the Delta; reducing Delta salinity to limit infestations of a marine borer, *Teredo*, which threatened wooden docks and structures; and improving navigation. Early environmental concerns were limited largely to fish passage and pollution from sewage. But the stability and strength of island levees have been a continuous concern, as have the costs of Delta management alternatives and the question of who should pay for them (Jackson and Paterson, 1977).

Today, California has an economy and society that could have only been dreamed of at the time of the earliest Delta studies. Although we retain many of the same concerns for the Delta, there have been changes in emphasis. New technology and infrastructure have eliminated the need to manage *Teredo* infestations (San Francisco Bay's first invasive species problem), but other alien invaders pose serious threats to California ecosystems, and society now places a higher value on maintaining a variety of aquatic and terrestrial species that depend on Delta habitats. In addition, greater reliance on the Delta for water supply and increased urbanization have heightened concerns about Delta water quality and about the weak levees that surround many Delta islands.

Some of these concerns will continue to evolve as a result of changing conditions in the Delta. As described in Chapter 3, increasing sea level rise, continued land subsidence, regional climate change, and increasing urbanization all contribute to the unsustainability of current Delta management. As California's population continues to grow, it is also likely that society will increasingly emphasize Delta services, including fish and wildlife habitat, recreation, urban housing, and water quality, making the Delta an even more important resource than it is today.

Any long-term management alternative for the Delta should be evaluated by its ability to address a broad range of concerns. In this chapter, we perform an initial evaluation of the nine alternatives described in Chapter 7. We first examine how responsive each alternative is to key Delta problems and concerns. We then evaluate, as best we can, how well different alternatives are likely to perform in terms of these concerns. Our aim is not to pinpoint "the" optimal solution but rather to identify several broad Delta alternatives with the most promise. Our analysis also serves to highlight the need for in-depth evaluation of the details of any Delta alternative before Californians make lasting policy decisions on the Delta's future.

Evaluation of Strategic Directions

A simple way to begin is to identify the major Delta issues that any alternative must address and to note how many of these issues each alternative is able to handle (Table 8.1). We have highlighted six issues likely to be important for key Delta interests:

- **Island flooding.** Does the alternative address long-term risks to Delta water supply, water quality, and land use from island flooding?
- **Water export quality.** Does the alternative provide a way to maintain or improve the quality of water exported to users south and west of the Delta?
- **In-Delta water quality for agricultural and urban users.** Would the alternative keep salinity levels sufficiently low to permit irrigation and urban water uses in at least parts of the Delta?
- **Water supply reliability.** Does the alternative provide a way to enhance the reliability of water supplies for Delta exporters?
- **Desirable species.** Does the alternative improve conditions for desirable fish and terrestrial species that depend on the Delta?
- **Urbanization.** Does the alternative provide sufficiently high levels of flood protection (exceeding 200-year average recurrence) and water quality to support urbanization in some parts of the Delta?

Table 8.1

Problems Addressed by Evaluated Alternatives

Alternatives	Island Flooding	Water Export Quality	In-Delta Water Quality	Water Supply Reliability	Habitats Supporting Desirable Species	Urbanization in the Delta
Freshwater Delta						
1. Levees as Usual	?					
2. Fortress Delta	X	X	X	X		X
3. Seaward Saltwater Barrier		X	X	X		
Fluctuating Delta						
4. Peripheral Canal Plus	X	X	?	X	?	?
5. South Delta Restoration Aqueduct	X	X	X	X	?	?
6. Armored-Island Aqueduct	X	X	X	X	?	X
Reduced-Exports Delta						
7. Opportunistic Delta	X	?	?	?	X	?
8. Eco-Delta	X	?	?	?	X	
9. Abandoned Delta	X				?	

NOTES: "X" indicates that the problem is addressed by the alternative; "?" means that it might be addressed, depending on details; cells left blank indicate that the alternative does not effectively address the problem.

In selecting these issues, we acknowledge that none of the alternatives will be able to address all of them entirely. In particular, we do not consider it feasible to eliminate or substantially reduce the risk of flooding for all Delta islands. Over the long term, some agricultural land will therefore go out of production. In our analysis, the key criterion for the feasibility of Delta agriculture is the extent to which an alternative provides adequate in-Delta water quality to maintain profitability on islands that do not flood.

Some alternatives respond to only a few concerns, whereas others respond to a wider range of problems. The Freshwater Delta alternatives do not look particularly promising in terms of their scope. If not combined with other alternatives, the Levees as Usual option (#1) looks particularly poor from all perspectives, because it is not designed to address any major problem over the long term. The Seaward Saltwater Barrier alternative (#3) also looks unpromising, because it is unable to solve many contemporary problems: It does not address environmental concerns and it makes urbanization more difficult. In effect, although it eliminates the need to maintain islands to keep the Delta fresh, it could increase flood risks.[1] Although the Fortress Delta alternative (#2) better protects many Delta islands, it, too, is unable to address environmental issues in the Delta. The maintenance of a freshwater system in the Delta does not permit the restoration of fluctuating salinity, which would facilitate the control of invasive species now threatening the survival of some key species.

All three of the Fluctuating Delta alternatives appear to have the potential to address most, and perhaps all, of the problems identified. For the two alternatives that contain versions of the peripheral canal, this potential depends on the details of canal design and implementation. Both canal versions address the risks of island flooding, in terms of water exports, by circumventing the Delta. The South Delta Restoration Aqueduct alternative (#5) also directly addresses water quality in the southern and eastern portions of the Delta. The ability of the Peripheral Canal Plus

---

[1] If a barrier is operated to keep water levels in the Delta higher than at present, it would worsen flooding risks, especially from spontaneous levee failures. For big flood events, it might perform a little better than other options, because it could reduce high tide effects for brief periods.

alternative (#4) to ensure water quality in the Delta, species protection, and urbanization depends on the extent to which complementary investments are made within the Delta. The compatibility of the South Delta Restoration Aqueduct alternative (#5) with some urbanization and with the restoration of delta smelt and other desirable species also depends on the details. The Armored-Island Aqueduct alternative (#6) is a type of through-Delta canal (rather than a peripheral one) but probably more porous on the east side and more fortified on the west side to allow managed salinity fluctuations to the west. It would tend to concentrate freshwater inflows in the eastern Delta and would fortify and protect some islands.

The Reduced-Exports alternatives—all of which are based on major changes in water export regimes—offer very different degrees of relief to Delta problems. As Chapter 6 indicates, users of Delta waters have some ability to adapt to changes in Delta exports, although the costs of certain adjustments are substantial. As we have envisioned it, the Opportunistic Delta alternative (#7) has the potential to address both ecosystem problems and the concerns of water exporters, but it anticipates a phase-out of some current land uses in at least parts of the Delta. The Eco-Delta alternative (#8) is essentially a variant of the Opportunistic Delta alternative that focuses on ecosystem needs. However, it offers the potential to satisfy some exporter concerns (both quality and supply reliability) as well as to address water quality concerns (particularly for environmentally friendly Delta agriculture). The Abandoned Delta alternative (#9) assumes a staged retreat from all Delta water and land uses, including environmental restoration. It therefore resolves the problem of island flooding by eliminating the need for Delta water supplies and economic land use. There could nevertheless be some ecosystem benefits to this alternative, resulting from its ability to increase salinity fluctuation in the western Delta and Suisun Marsh area.

Performance Criteria and Likely Performance

Of course, Table 8.1 does not indicate performance—or how well each issue would be addressed by each alternative. A major study of solutions for the Delta, drawing on a finite set of detailed performance criteria, would be needed to provide such an evaluation. In this initial evaluation, we take a much simpler approach. Using available information, we provide our best judgment on how well each alternative is likely to stack up across three broad criteria: environmental, water supply, and economic performance (Table 8.2). This analysis does not include the full range of current objectives for the Delta; there will inevitably be some controversy regarding any selection of evaluation criteria and estimation of performance. Nevertheless, this analysis offers some guidance on favorable directions to take. It also illustrates the type of comparative analysis that is desirable for long-term infrastructure decisionmaking. The following provides a brief outline of our three major performance criteria.

Environmental Performance

Under current law, environmental performance is an overriding concern for Delta management, because all users must consider the effects of their actions on endangered and threatened species. Our assessment of environmental performance is based on our judgment of how well each alternative could be adapted to improve the health of Delta-dependent desirable species; this evaluation is based on the understanding of the Delta ecosystem discussed in Chapter 4. One aspect of environmental performance is the entrainment of fish and fish larvae by export pumps. Available information is not sufficient to evaluate this problem thoroughly, but it is likely that any through-Delta alternative, as well as some peripheral canal alternatives, would need to include components to limit fish entrainment. A variety of options exist to mitigate this effect, including changing various intake locations, altering pumping patterns, and employing finer fish screens or bank filtration. Options are likely to vary in effectiveness and cost.

Water Supply

Our evaluation of water supply performance focuses on the ability of each alternative to provide water exports of sufficient quality to points south and west of the Delta. Table 8.2 summarizes this assessment in terms of volumes available in typical years. This evaluation draws on numerous water management studies, including the CALVIN model results presented in Chapter 6 and elsewhere (Jenkins et al., 2001, 2004; Lund et al., 2003; Tanaka and Lund, 2003; Tanaka et al., 2006; Medellin et al., 2006), and various results from the water resources simulation model (CALSIM) (Department of Water Resources, 2006; U.S. Bureau of Reclamation,

2005).[2] Although the different studies' methods and assumptions lead to a variety of results, they permit an assessment of the alternatives that seem most promising for water supply purposes. For water supply for agricultural and urban users within the Delta (a function of water quality in the Delta), we are currently unable to go beyond the qualitative assessment provided in Table 8.1.

Economic Performance

Economic performance relates to the diverse set of costs associated with each alternative. Costs include not only new investments and operating expenses but also the direct and secondary economic effects from changes in the availability of Delta land and water services. Investment costs may be incurred for new water supply facilities, improved levees to protect infrastructure and buildings from floods, gates, barriers, fish screens, and other infrastructure. Operating expenses arise from pumping, treatment, and maintenance costs as well as from levee repair and recovery costs from levee failures. Changes in service availability include costs from changes in water scarcity and reliability as well as from changes in water quality. As shown in Chapter 6, with foresight and preparation the California economy has significant potential to adjust, at some cost and institutional inconvenience, even to extreme policy changes in Delta exports. Land use transitions are also possible, including modifications of activities that now rely on current Delta levees. A key question is whether alternatives that seek to avoid major adjustment costs are preferable overall to those with major changes. Because these various costs would be borne by different groups and regions, questions of fairness will be an inevitable part of this policy discussion, in addition to the overall costs. Possible mitigating actions are discussed in Chapter 9.

Here, we provide some rough comparisons for illustrative purposes, focusing primarily on investment costs and adjustment costs for water users. An in-depth analysis of alternatives would need to consider a wider range of costs, including adjustment costs for users of other civil infrastructure and secondary economic effects. We estimate investment

---

[2]CALSIM is DWR's and USBR's model of CVP and SWP operations and deliveries. This model is widely used to evaluate water deliveries and operations of these major water projects.

Table 8.2
Comparisons of Likely Performance

Alternatives	Environmental Performance	Water Supply Exports	Economic and Financial Costs
Freshwater Delta			
1. Levees as Usual—current or increased effort	Unpredictable but, if present trends continue, poor	6+ maf/year short term; 0–6+ maf/year long term	~ $2 billion, plus increasing costs of failure and replacement
2. Fortress Delta (Dutch standards)	Poor, favors low productivity habitat mainly for alien species		> $4 billion
3. Seaward Saltwater Barrier	Creates a freshwater system favoring warm-water alien species	6+ maf/year long term	$2 billion–$3 billion
Fluctuating Delta			
4. Peripheral Canal Plus	Depends on configuration of canal but potentially allows more natural flow regime through Delta, favoring desirable fish	6+ maf/year long term	$2 billion–$3 billion; < $70 million/year additional water scarcity costs
5. South Delta Restoration Aqueduct	Provides a variety of habitats in South Delta; effect on desirable species depends on configuration and operation	6+ maf/year long term; lower export water quality than for Peripheral Canal Plus	$2 billion–$3 billion; < $41 million/year additional water scarcity costs
6. Armored-Island Aqueduct	Provides a variety of habitats but hard on anadromous fish (e.g., salmon)		$1 billion–$2 billion+; < $30 million/year additional water scarcity costs

Table 8.2 (continued)

Alternatives	Environmental Performance	Water Supply Exports	Economic and Financial Costs
Reduced-Exports Delta			
7. Opportunistic Delta	Provides opportunity for habitat diversity but could resemble Alternative #1	Highly variable, 2–8 maf/year	$0.7 billion–$2.2 billion in Delta and near Delta facilities; water scarcity cost < $170 million/year
8. Eco-Delta	Good, because system managed to favor desirable species	Highly variable, probably 1–5 maf/year	Several billion dollars (eco-restoration only); + water user investments and water scarcity cost < $600 million/year
9. Abandoned Delta	Probably poor, favoring alien species, similar to Alternative #1	0	~ $500 million additional capital costs plus ~ $1.2 billion/year scarcity and operating costs

NOTES: Unless specified, costs listed are for capital investments (see Appendix E). All alternatives except #9 (and possibly #2) would require additional investments for urban levees to provide flood protection exceeding 200-year average recurrence. All alternatives except #8 and #9 would require additional investments for ecosystem restoration. All alternatives foresee losses in Delta farm revenues, but these are included above only as part of water scarcity cost estimates for options #4 through #9. For options #1 to #3, there would be additional losses from land going out of production from island flooding. Adding finer fish screens or bank filtration to intakes to reduce fish and larvae entrainment would increase costs and potentially reduce pumping capabilities.

costs by using various published and unpublished sources and water user adjustment costs by drawing on the CALVIN and DAP results presented in Chapter 6. The Eco-Delta alternative is the only alternative that explicitly provides investment cost estimates for ecosystem restoration; these should be viewed as an upper bound on such investments, at least some of which would accompany some of the other scenarios. Because the trajectory of urbanization in the Delta may vary, we do not include the additional costs of urban levee fortification that would be necessary to accommodate such growth. These costs are likely to run in the range of $200 million to over $1.5 billion if 100–150 miles of levees must be upgraded for urban development. Additional levee costs might be incurred to protect civil infrastructure on interior islands. However, some levee investments in the Fortress Delta alternative could double as protection for urban areas and infrastructure, depending on the location of urban settlements and infrastructure networks relative to levees that need to be enhanced to protect Delta water supplies. Finally, we do not incorporate the costs of a mitigation program to ease adjustment for those bearing particularly high costs under the various alternatives (although for water users in the Delta, the estimated adjustment costs provide some indication). Detailed cost estimates for each alternative are discussed in Appendix E.

Summary Evaluation of Alternatives

Our judgment of the overall promise of each alternative appears in Table 8.3. Our analysis suggests alternatives that should be eliminated from further consideration and those that merit further exploration and refinement. The table also provides a thumbnail rationale for each of these judgments, which we expand on below.

Freshwater Delta Alternatives

On all counts, the three freshwater alternatives appear unpromising.

Perpetuating the Delta as a homogeneous freshwater body would be environmentally damaging. This strategy fosters the wrong kinds of habitat for native species and tends to promote undesirable invasive species.

Table 8.3

Summary Evaluation of Alternatives

Alternatives	Summary Evaluation	Rationale
Freshwater Delta		
1. Levees as Usual—current or increased effort	Eliminate	Current and foreseeable investments at best continue a risky situation; other "soft landing" approaches are more promising; not sustainable in any sense
2. Fortress Delta (Dutch standards)	Eliminate	Great expense; unable to resolve important ecosystem issues
3. Seaward Saltwater barrier	Eliminate	Great expense; profoundly undesirable ecosystem performance; water quality risks
Fluctuating Delta		
4. Peripheral Canal Plus	Consider	Environmental performance uncertain but promising; good water export reliability; large capital investment
5. South Delta Restoration Aqueduct	Consider	Environmental performance uncertain but more adaptable than Peripheral Canal Plus; water delivery promising for exports and in-Delta uses; large capital investment
6. Armored-Island Aqueduct	Consider	Environmental performance likely poor unless carefully designed; water delivery promising; large capital investment
Reduced-Exports Delta		
7. Opportunistic Delta	Consider	Expenses and risks shift to importing areas; relatively low capital investment; environmental effectiveness unclear
8. Eco-Delta	Consider	Initial costs likely to be very high; long-term benefits potentially high if Delta becomes park/open space/endangered species refuge
9. Abandoned Delta	Eliminate	Poor overall economic performance; southern Delta water quality problems; like Alternative #1, without benefits

Environmental performance would be worst with the Seaward Saltwater Barrier option, because it would also obstruct fish passage between the bay and the Delta.

Water supply performance would be good in the Fortress Delta and Seaward Saltwater Barrier alternatives—about 6+ maf per year of export deliveries (comparable to recent export levels). The exception is the Levees as Usual alternative, in which deliveries would be likely to decrease significantly as episodes of levee failure increase. Land subsidence and sea level rise make the Levees as Usual option increasingly unreliable and risky for water supplies. The Seaward Saltwater Barrier would be useful in maintaining a freshwater Delta after multiple island failures from a major earthquake and thus may be more dependable than other freshwater options in terms of water supply. But its structure and gate mechanisms would also be severely challenged by seismic events, when they are likely to be most needed.

Finally, the Freshwater Delta alternatives tend to be relatively expensive because all three are based on major levee or barrier investments. Investment costs for these options range from approximately $2 billion for Levees as Usual to over $4 billion for a Fortress Delta; costs for the Seaward Saltwater Barrier probably lie somewhere in between. Additional ongoing costs for levee maintenance and repair would be required for all these alternatives. Levees as Usual would have comparatively low initial capital costs but increasingly high costs of upkeep.[3] Costs for levee repair and levee failures would be particularly large and frequent. Additional failure recovery costs under this alternative could average on the order of $100 million per year.[4] The Fortress Delta alternative is likely to entail high investment costs as well as high ongoing maintenance and upkeep, given the increasing pressures of flood flows, sea level rise, and seismic risk that will face the Delta in the years ahead. However, failure recovery costs under this alternative could be considerably lower than those under Levees as Usual. Failure recovery costs also could be substantial for a Seaward Saltwater Barrier option, if Delta islands were maintained once the water supply risk had been eliminated.

---

[3]For instance, DWR estimates that repairs to weakened or failed project levees currently cost on the order of $5,000 per foot ($28 million per mile).

[4]Estimated on the basis of a failure cost of roughly $10 billion, with a probability of failure of 1 percent per year. Such rough estimates could be refined using results from the ongoing DRMS. Even this relatively low estimate implies a present value of failure recovery costs of $2 billion (roughly the initial capital cost), and it does not include additional catastrophic event costs faced by state and local governments.

Overall, these solutions perform poorly environmentally, do not appear to offer cost-effective long-term solutions to water supply issues, and would be relatively expensive to carry out and maintain. We recommend eliminating all three of these alternatives from further consideration.

Fluctuating Delta Alternatives

Each of the Fluctuating Delta alternatives is promising for our three performance categories. Of course, the degree of favorable performance for any of these alternatives would depend greatly on the details of operation and implementation.

Environmentally, the Fluctuating Delta alternatives seek to break the dependency of the Delta on water exports. The Peripheral Canal Plus and the South Delta Restoration Aqueduct would do so by circumventing the Delta, whereas the Armored-Island Aqueduct would reconstruct through-Delta conveyance so that water export flows are largely isolated from the western part of the Delta, where salinity could fluctuate. These alternatives are likely to have good environmental performance, as they would provide a wide range of environmental habitats to support desirable species and offer greater patterns of fluctuation, which inhibit many potential and current invasive species. Their detailed environmental performances would differ with the particulars of each alternative.

Water supply export performance is also quite good for all three alternatives, with volumes in the range of 6+ maf per year. Exports are limited mostly by the capacity of downstream conveyance capacity and upstream water availability and depend much less on Delta conditions than at present, although enough fresh water would still need to flow into the Delta to maintain desired salinity fluctuations. Compared with the current through-Delta conveyance system, the Peripheral Canal Plus would enhance export water quality, because it avoids blending higher-quality Sacramento River water with the lower-quality water of the Delta. The reliability of these alternatives should be greater for floods, earthquakes, other Delta island failures, and many risks to water exports associated with protection of aquatic species.

Significant capital costs would be required for all three of these alternatives, although the costs presented here are highly uncertain. There would be some additional pumping costs for the Peripheral Canal Plus and

South Delta Restoration Aqueduct alternatives. Water scarcity costs would arise from lost agricultural production on some Delta islands, which would result from increased salinity levels necessary to support habitat favorable to desirable species. Given some likely improvement in water export reliability, water scarcity costs south and west of the Delta might decrease compared to current conditions but probably by no more than $20 million per year on average.

Fluctuating Delta alternatives would potentially improve the Delta's environment and its water export reliability and quality. The economic cost of each would be considerable but probably less than most of the freshwater alternatives. Perhaps most important, given the variety of changes facing the Delta, these alternatives tend to add flexibility to the system and to provide greater adaptability to changes in future conditions. We recommend that all three Fluctuating Delta alternatives be given further consideration.

Reduced-Exports Alternatives

The three Reduced-Exports alternatives rely on various modifications of Delta export pumping; our performance criteria indicate mixed potential.

The environmental performance of these options differs with the degree of pumping changes required to introduce greater habitat variability and specialization into the Delta. Of course, the details of environmental performance would differ with implementation details. It is interesting to note that abandoning the Delta, without any restoration actions, leads to a generally unfavorable long-term environmental condition similar to that of the Levees as Usual alternative. Any additional salinity fluctuation introduced here would be much less productive without other environmental restoration actions.

In the two alternatives in which water exports are curtailed rather than eliminated—the Opportunistic Delta and the Eco-Delta—exports would become more variable than they are currently. Although neither of these alternatives relies on significant new water supply infrastructure, investment costs remain substantial. Opportunistic pumping would probably be accompanied by some off-stream storage near the pumps to provide the flexibility to pump more water during high flow periods than can be accommodated by existing canal capacity. By contrast, the Abandoned

Delta has fairly low capital costs (mainly for strengthening interties) but very high operating and water scarcity costs.

Our evaluation of this set of alternatives finds that two merit further consideration. The Opportunistic Delta and the Eco-Delta—both of which encourage habitats supportive of desirable species in the Delta without constructing a peripheral or through-Delta canal—are worth considering further. Both provide the potential for better management of the Delta environment while permitting continued use of the Delta for other purposes, including water exports (albeit at reduced levels). By contrast, we do not consider it worthwhile to further consider the Abandoned Delta. The water supply and scarcity costs of this approach are unreasonably high and accompanied by likely serious salinity problems in the southern Delta as well as poor environmental performance for native species. Sea level rise and climate warming would likely accelerate the deterioration of the Delta if it were abandoned. And abandoning the Delta also reduces the environmental, land, and water resources available to California for adapting to climatic change, including the ability to move water to areas where it creates more economic well-being.

Desirable Characteristics of a Delta Solution

This analysis points to some of the characteristics that would be desirable to include in any Delta solutions.

Hybrid Solutions

To address most Delta problems, any comprehensive solution will need to contain a hybrid of several strategies. For example, a peripheral canal on its own might address some problems, but it leaves many others unaddressed. Likewise, levees will be an important part of any Delta solution, but levees alone are likely to be disastrous for some objectives and economically unreasonable overall. Although the recently passed bond measures provide valuable support to flood protection in the Delta, the mere funding of levee construction and reinforcement alone will be insufficient; more profound and integrated redesign of the system will be needed. Both in the comparison of the problem addressed by each alternative (Table 8.1) and in the summary evaluation of alternatives (Table 8.3), the more promising approaches tend to contain hybrid solutions.

"Soft Landing" for the Delta

A major motivation for changing management of the Delta is the increasingly fragile nature of the current Delta's environmental, land use, and water supply functions. There is an unacceptable probability that the Delta's current management and services could abruptly crash in ways that would be catastrophic environmentally and economically. Most of the alternatives considered here seek a soft landing from the Delta's current severe disequilibrium and vulnerability. Efforts to address short-term emergencies and failures in the near term are necessary (as the DRMS is attempting to explore), but longer-term efforts should be dedicated to preventing such failures and catastrophes and should significantly alter the Delta from its increasingly unsustainable form.

Trial Solutions

Broadly obvious and ideal solutions do not exist for the Delta's problems. All promising solutions entail significant uncertainties. The implementation of any promising solution should involve some experimentation before making irreversible decisions, to limit the extent of failures. However, the Delta is not a science experiment. Performing some field experiments may sometimes be desirable to provide timely information to help improve management, but such experiments cannot provide absolute certainty and should not be used as a strategy to delay decisions. Computer modeling is another form of experimentation, based on mathematical representations of our current knowledge. In some cases, trial or modeled solutions should allow us to accelerate decisionmaking by making small experimental decisions in the field or in computerized settings. The original forms of adaptive management (Hollings, 1978) envisioned a close relationship among computer model development, field experiments, and management policies over time. However, the urgency of the Delta's problems probably will not permit casual, nonaggressive forms of adaptive management to be successful. Only more aggressive forms of adaptive management are likely to succeed in developing understanding and management approaches in time to preserve species that are now severely at risk.

Phased Implementation

The instantaneous implementation of a complete solution package is unlikely. Any solution is likely to require too much capital to be implemented all at once, and there will most likely be too many uncertainties and controversies to address in the course of implementation. For these reasons, phased implementation is likely to be both necessary and desirable. Phased implementation can take two forms: (1) planned phased implementation, in which the details in a phase are scheduled and orchestrated, and (2) opportunistic implementation, in which events in the Delta provide opportunities to make desirable changes relatively easily. An example of this second type would be failure of a levee on an island that is scheduled to become open water habitat or a floodway. Such a failure would present an opportunity to accelerate this phase of a long-term plan. To take advantage of such opportunities, it would be helpful to develop a "do not resuscitate" list of nonstrategic Delta islands, as described below. Phased implementation would also allow us to make progress and establish strategic direction, while adapting the strategy as uncertainties become better understood.

Stakeholder Involvement in Implementation and Operations

The many functions of the Delta are operationally complex. One concrete accomplishment of the CALFED process has been improved operational communication and coordination among various interests regarding Delta water management activities. Communication and coordination will be desirable features for the operation of any future Delta alternative. The many parties interested in the Delta have expertise and resources that are unavailable to the state and federal agencies that are charged with developing and implementing solutions. Local reclamation districts are probably the best experts on current levees; similarly, local developers and city officials know a great deal about urban land potential; and water contractors know the most about achieving water quality goals for their customers. This is not to say that the solutions to the Delta's problems are likely to be developed purely by consensus, given the inevitable tradeoffs involved. But local expertise should be involved to improve the design and implementation of Delta solutions. Centralized and isolated crafting of solutions to complex local problems is unlikely to be successful.

Reducing and Managing Uncertainties

Although our knowledge about some key drivers of change in the Delta has increased greatly in recent years, some major uncertainties still may affect the viability or design of different Delta alternatives. There is also considerable uncertainty as to how various alternatives would affect ecosystem performance, water supply and quality reliability, and other objectives. As part of any exercise to craft detailed long-term solutions, investigations will be needed into these areas. These investigations may include problem-oriented modeling and laboratory analysis as well as field experimentation. To be useful, investigations will need to be conducted in a coordinated manner.

- **Climate change.** To date, we have a general understanding of the effects of climate warming on the Delta. Faster melting of the Sierra Nevada snow pack is likely to increase the risk of flood events, and sea level rise is expected to raise pressures on Delta levees (see Chapter 3). Although we know that sea level rise could increase western Delta salinity under current operations (Department of Water Resources, 2006), we know relatively little about the effects on salinity under different operational scenarios. Hydrodynamic modeling studies are beginning to explore such effects. Research is also needed to help clarify how changes in water temperature will affect the distribution and abundance of some native and alien species, including delta smelt, striped bass, and overbite clam.
- **Alien species.** Given the dominance of alien species within the Delta, finding management techniques to discourage alien invaders and to encourage the few remaining native species is a major challenge. There are important gaps in our knowledge of the response of existing alien species to salinity, velocity, water clarity, and other manageable aspects of physical habitat. Short-term research efforts can help assess viable management solutions. Policy actions (discussed below) will be needed to help stem the arrival and establishment of new invasive species.

- **Runoff contamination.** Many investigations have concluded that spikes in contaminated runoff from agricultural and urban areas may be an important contributor to the decline in open-water fish species such as the delta smelt (Dileanis, Bennett, and Domagalski, 2002). Regulations are being introduced, but this process is slow and politically difficult. Knowing more about runoff and its effects will assist in environmental planning and policy implementation for both land and water uses.
- **Urbanization.** Although the general context of urbanization pressures in the Delta is well understood, there is as yet no clear understanding of the extent to which development in the Delta is compatible with environmental sustainability and no overall analysis of its implications for flood risks. Should urbanization be directed away from some areas or guided by special subdivision and building regulations in some others? How should flood control and local drainage be managed for these areas?
- **Recreation.** There is an urgent need to better understand the scale and scope of current and potential recreational uses of the Delta. The Delta is already an important recreational resource. As the region's population grows, it is quite likely that the economic benefits of recreation will overshadow those of traditional agriculture, if it does not already do so.[5]
- **Failure recovery costs.** Many of the Delta alternatives have a significant probability and cost of failure, from levee failure or other causes. These costs and probabilities should be assessed to serve as contributions to the development and comparison of alternatives. The current DRMS effort is providing useful work in this regard for island levee failures under current conditions (www.drms.water.ca.gov; Jack R. Benjamin and Associates, 2005).
- **Ecosystem research.** As discussed in Chapter 4, a variety of directed research is needed to more precisely and accurately define the habitat needs of key species and inform the acquisition and management of many particular habitats and locations.

---

[5]The long-term potential of recreation was highlighted at a workshop on Delta land use organized by a group of landscape architects from UC Berkeley and the Natural Heritage Institute in March 2006.

- **In-Delta land use and habitat.** Although our analysis suggests that the local specialization of island uses and the allowance of fluctuating salinity within the Delta offer many advantages, there is as yet limited knowledge of the best environmental and economic uses for individual islands and other peripheral areas. Such information is essential to assess the costs and benefits of managing the Delta through local specialization. Habitat plans that incorporate contingencies and uncertainty will better allow us to learn, adapt, and take advantage of opportunities.

All major uncertainties cannot be resolved before decisions on the Delta should be made. But not all issues are critical to all decisions. A successful long-term strategy should have a consistent general approach. Some components can be undertaken quickly or in stages with little uncertainty, whereas others can be delayed until there is greater clarity (but probably not perfect certainty). And some components will need to be experimental in nature.

The greatest error would be to wait and make decisions only when all uncertainties have been eliminated. There is cost and considerable risk from inaction and indecision, and action must be taken before dire events unfold. Many important decisions and directions can and must be established with existing scientific and technical understanding of the Delta and its uses. Uncertainty can rarely be eliminated; it must always be managed.

Crafting, Evaluating, and Gathering Support for Better Alternatives

Though preliminary, the evaluations presented here provide some insight into what kind of alternatives for managing the Delta would be desirable or undesirable overall. Moreover, the approach we have taken—to explicitly evaluate stated alternatives on a range of performance objectives—is a rational and promising way to arrive at an alternative that will function well on the ground. But our analysis is neglectful in three ways. Technically, our effort was too limited in time and resources to consider detailed operational plans or to conduct in-depth evaluations. Second, given the limited scope of this work, we were unable to examine a wider range of hybrid alternatives. Nevertheless, we believe that our analysis provides a good coarse filter for winnowing out unpromising approaches and for introducing promising ideas into ongoing discussions. Third, politically, our analysis is purposefully naïve. No good technical solution is likely to be implemented without political support. But the converse is also true. On its own, a political process will not be able to develop new technical alternatives or provide a technically sound analysis of alternatives. A careful and disinterested technical process—at arm's length from the political process—will be essential for crafting a viable future for the Delta.

Basing Solutions on Improved and Integrated Understanding

Developing and evaluating solutions for the Delta's complex problems will require a technical synthesis of existing and new information across a wide range of Delta-related subjects and perspectives. Such synthesis is most transparent, rigorous, and effective if conducted with the explicit aid of computer models (California Water and Environment Modeling Forum, 2005). To make results more reliable and insightful, quality control and visualization tools are important aspects of this synthesis. Despite significant investments in scientific and technical tools, the scientific and policy communities have neglected the development and testing of models and data that integrate the many aspects of ecosystem functioning, water supply and quality, and land use that determine the viability of various Delta services. The CALVIN and DAP models applied in Chapter 6 are primitive examples of what can and should be accomplished in this regard. Many models for hydrodynamics and water quality (DSM2, FDM, etc.), operations planning (CALSIM), and economics (CALAG and LCPSIM) also exist and should have important roles. To date, none of these models are entirely suited to the types of studies needed to map out long-term futures for the Delta. Models of land use and habitat in the Delta (perhaps expanding on DAP) would provide a basis for integrating land, water, and habitat decisions for the Delta. It is necessary to prepare a technical basis for exploring, developing, and comparing detailed Delta alternatives.

A combination of basic and applied research also will be required to address or narrow some of the major uncertainties noted above. Most of this research should be developed within a solution-oriented framework, as opposed to using an exploratory, basic science approach. Although our understanding of the Delta's complex problem will never be perfect, the scientific and policy communities have not made the most of integrating what we do know and have not always focused research efforts strategically on the most important questions. By developing and documenting an integrated understanding of the Delta, we will have an unprecedented ability to develop and test potential solutions and provide greater scientific assurance that taxpayer and stakeholder resources are being effectively employed.

Short-Term Actions

Solving the Delta's problems cannot occur quickly, even if action begins immediately. Developing and implementing a deliberative and thoughtful solution to this long-term problem will require years rather than months. In the face of this long-term strategic decision for California, prudence suggests several short-term actions:

- **Establish emergency-response and preparedness plans.** Levee failures are likely to occur at any time, as illustrated by the failure of the Lower Jones Tract levee in June 2004. Federal, state, and local agencies need to be prepared for large and small failures on short notice. The state and many local agencies have realized this problem and are taking useful steps. For water agencies that rely on Delta water, necessary measures include developing extended water export outage plans. With measures such as regional interties, water sharing agreements, local supply development, and drought contingency plans, the costs of losing a year of Delta exports can be reduced by a factor of 10 (Chapter 6). Other infrastructure providers that rely on the Delta, such as Caltrans, the railroads, and power companies, need similar contingency plans and should consider making new investments so that their networks are less susceptible to levee failure (for instance, burying pipelines or repositioning stretches of road). Creating a program for the rapid repair of critical levees—such as the one launched in 2006—and emergency flood response plans are also urgent.
- **Create a "do not resuscitate" list of Delta islands.** To safeguard the state's financial resources and force some movement toward a

long-term solution, the state should create a "do not resuscitate" list of Delta islands that do not have strategic value in terms of homes, infrastructure, or water supply. When these islands fail, the state would not intervene. It is already apparent that preserving or rebuilding levees for some islands is not in the state's interest (Logan, 1990). This is an important policy decision that would provide important financial savings. As noted above, it would also facilitate experimentation with environmental uses of flooded islands for habitat and flood bypasses.

- **Provide protection for urbanizing areas.** One of the few drivers of change in the Delta that we can affect is urbanization. Once established, however, urbanization of land is essentially irreversible. There is a need to protect existing urban development, but urbanization should not occur in locations that cannot or will not be protected from flooding. Local land use controls have not always been sufficient in this regard. New development projects should not impose irresponsible levels of risk on local residents and state and local governments. Habitat of particular value to Delta species should be acquired through purchases or set-asides (see Chapter 9).

- **Prevent the introduction of new invasive species.** In addition to existing problems with alien species, the Delta faces the continual threat of the arrival of new species, which can upset whatever balance has been achieved with previous invaders. Risk reduction can be accomplished through better regulation of known sources of alien species (e.g., ballast water and the aquarium trade) and better preparation to eradicate new invaders before they spread (e.g., northern pike). There is also a need for emergency response and preparedness for new invasions; rapid eradication of an invader while it is still localized can prevent future problems.

- **Initiate a technical solution effort.** A coherent and substantial effort currently does not exist for identifying, exploring, developing, and evaluating promising long-term technical solutions for the Delta. This effort will require development of data, modeling, and visualization tools to form the foundation of technical studies and to provide assurances and the communication of results for policymaking. A solution-oriented science program also is needed.

This technical effort will be a necessary part of any process to find and implement an effective long-term solution for the Delta.

- **Focus on Suisun Marsh and Cache Slough.** These two large areas have favorable prospects, from an ecosystem perspective, under various Delta change scenarios. Studies should begin immediately to model the likely future effects of the major drivers of change in these areas and to suggest how these effects can be managed to favor desirable organisms and hydrologic pathways. Land acquisition or easements should begin immediately, from willing sellers, in areas that are most likely to be affected by flooding and levee failure or to be beneficial to desired species. Planning and management efforts that are already under way in both areas should be enhanced to improve landowner and stakeholder understanding of alternative futures. Similar efforts should be undertaken in other areas peripheral to the central Delta, such as the Cosumnes River area, that are or have the potential to become centers of abundance for desirable species.

- **Begin discussions of governance and finance.** Technical studies are likely to require several years to complete. Discussions and agreement on the governance and finance of any Delta solution will likely take at least as long and involve at least as many difficulties. Such discussions should begin soon. Technical, political, and financial work all need to occur simultaneously, although not always in the same room. Having some distance between the political and technical processes provides state and federal elected officials with greater assurance that final proposals have received both stakeholder and technical scrutiny and evaluation. In the next chapter, we provide some thoughts on how to move forward in developing financial and governance options for the Delta.

On its own, a stakeholder- or policy-driven process is unlikely to generate functional long-term solutions to the Delta's problems. For this reason, a serious, systematic technical effort, which has been largely absent in the recent past, will need to accompany exercises such as the Delta Vision effort. Such a technical effort can and should enrich policy discussions by suggesting promising new alternatives, deterring unproductive discussion of unpromising alternatives, and providing voters and elected officials with greater confidence and information on the costs, benefits, and likely tradeoffs of alternative solutions.

9. Financing and Governing a Soft Landing

"The best is the enemy of the good."

Voltaire

The Delta of the future cannot be all things to all people, as CALFED agreements seemed to promise. The current Delta is unsustainable, and all options for strengthening its long-term prospects involve tradeoffs. Alternatives that seek to maintain a freshwater Delta are not compatible with improved conditions for native species. Alternatives that allow for a fluctuating Delta could achieve this goal, but they would entail some loss of Delta farmland and would affect other Delta users. Alternatives involving reduced Delta exports, with only seasonal pumping, would generate losses for water exporters as well as for Delta farmers. Investments in any given Delta alternative also imply tradeoffs, because these resources could be used to fund other priorities.

The risks are very real that a desire to protect the status quo will prevent the adoption of approaches that achieve much better outcomes for California—i.e., those that generate the greatest benefits overall relative to the costs Californians are willing to support. Pressure to protect the status quo is likely to come from various quarters. For example, Delta farmers and other land-based interests could be expected to push to maintain a freshwater Delta in its current configuration, to continue existing and planned land uses. And although water exporters might be more open to other alternatives, they share an interest in keeping user contributions to a minimum and relying on taxpayer support, as under CALFED.

To move beyond the status quo, California will need to consider new approaches to financing a Delta solution. First, this means resuscitating and strengthening the CALFED "beneficiary pays" principle. With better ground rules for user contributions, it should be possible to channel available public funds to support the parts of an investment package that are truly public in nature. Second, this means devising mitigation packages to soften the costs of adjustment. Mitigation is a different approach from that pursued by CALFED; instead of insisting that a Delta solution can provide direct benefits to all stakeholders, it acknowledges that there will be losers as well as winners in any long-term Delta strategy. By compensating those who lose out, mitigation can create incentives to move beyond the status quo. In this chapter, we provide some initial thoughts on how a new funding approach for the Delta might work.

Developing a viable Delta solution will also require innovative approaches to governance. Although it is beyond the scope of this report to provide a detailed analysis of governance questions, we provide some thoughts on two central issues: improving coordination of land use in the Delta and providing better incentives to manage water resources.

Funding Principles for a Soft Landing

Financial considerations are central to creating a successful long-term strategy for the Delta, given the magnitude of the sums required and the extended time frame over which investments will need to be undertaken. By our rough estimates, water infrastructure costs alone are likely to be in the range of several billion dollars, with significant additional sums required for stronger urban levees, ecosystem restoration efforts, and adjustments by other infrastructure providers (see Chapter 8). It is not realistic to expect taxpayer dollars to meet all, or even most, of these costs, given other demands on public spending. For instance, the recently approved bond funds for flood control earmark some $3 billion to $4 billion for the entire Central Valley—a very large sum relative to past state contributions.[1] Yet there is likely to be considerable competition for these funds, because some of the most pressing flood risks are in heavily urbanized areas upstream of the Delta. Beyond this, general obligation bond financing of water supply infrastructure (repaid with general state tax revenues) establishes poor incentives for local water managers to operate efficiently. If someone else is paying, it is always easier to ask for more. Thus, it is necessary to consider other options.

---

[1]Proposition 1E earmarks $3 billion to flood control within the Central Valley and $300 million to other regions and it provides $790 million for flood protection activities without geographic restrictions. Proposition 84 allocates $275 million to flood control in the Delta and another $525 to flood protection activities without geographic restrictions.

The Beneficiary Pays Principle Is More Relevant Than Ever

One of CALFED's clear failures was its inability to mobilize adequate stakeholder contributions to its investment portfolio. Everyone had signed on to the "beneficiary pays" principle, but stakeholders tended to argue that program elements provided public benefits, and should therefore be funded by state and federal taxpayers. As noted in Chapter 5, considerable local funds were expended on local water supply investments—notably groundwater storage, water use efficiency, and recycling. But general obligation bonds were also used to support these investments—making it cheaper for water agencies to stretch or expand water supplies—on grounds that this lessened pressures on the Delta ecosystem. Water users who stood to gain from new surface storage investments made similar arguments, without offering to fund a significant share themselves. Today, many Delta stakeholders are calling for massive public investments in Delta levees, even though many of the beneficiaries are clearly private or localized in nature: water exporters, Delta farmers and land developers, power and rail companies, and users of the local road network.

Because the costs of any new Delta strategy are likely to well exceed the funds available from state and federal coffers, better ground rules on funding contributions are needed. User finance—that is, payment by the actual users of the investments—has many advantages. It frees public funds for truly public purposes, such as ecosystem restoration and mitigation, and it helps ensure that many investments are cost-effective. If water users are unwilling to finance investments that increase the reliability of their water supply, chances are that the investment is not a sound one. If landowners are unwilling to contribute to the costs of flood protection, chances are that the value of the land to be protected is too low to merit such investments.[2]

User contributions would be especially relevant for collective infrastructure investments in both water supply and flood protection. Water exporters should be expected to fund improvements in water supply

---

[2]Local levee investments will also be too low if someone else is liable for flood damages. Since the 2003 *Paterno* decision, the state has been held liable for damages in areas behind "project levees" belonging to the Central Valley flood control system, which includes some Delta levees.

reliability, and a variety of beneficiaries should be expected to contribute to programs to reduce flood risks. It is often argued that mobilizing user contributions to Delta flood control is too complex, given the many interests involved and the fact that some of them—such as Caltrans—lack specific budgets to pay for such programs. But straightforward precedents exist for user finance in other areas of public safety. For example, the private sector finances most investments in seismic retrofits and prevention; Bay Area bridge users pay a surcharge to help fund seismic retrofits of bridges. There is no reason why a beneficiary pays principle could not apply to infrastructure adjustments in the Delta. For new homes and businesses, developer fees are a straightforward way to collect up-front contributions to flood protection and property assessments can be used to cover maintenance costs. The key challenge is to ensure that these fees and assessments are high enough to cover the costs of building and maintaining adequate protection. If not, the local community (and state taxpayers) will be left footing the bill.

Apportioning Costs of Large Water Projects

For water supply investments large enough to require the participation of multiple parties, one stumbling block facing CALFED was the lack of agreement on how to apportion costs among beneficiaries: Should each water user be required to pay the same amount for each unit of water received, or might some sort of sliding scale be appropriate? This question is particularly relevant for Delta exports. In a typical year, agricultural water use employs most (72%) of the direct diversions from the Delta, yet most agricultural uses cannot justify costs as high as those urban users are willing to pay.

Two central problems facing any public project are how large to build it and how much to charge users to cover the costs of the project. Standard economic calculations of marginal cost pricing, whereby all users are charged the incremental project cost, typically will fail to recoup total costs of water projects because the incremental cost falls as the size of the project expands. These economies of scale occur because building water projects often involves a large fixed cost and a relatively small constant per unit operating cost. Thus, spreading the fixed costs over greater capacity lowers the incremental unit cost.

An analogy can be made with the cost of operating a passenger jet. The basic costs of operating the jet are largely the same regardless of the number of passengers. The incremental cost of a student in the back of the plane is little more than peanuts (the in-flight snack), so how much of the fixed cost of flying the plane should be charged to the student and how much should be charged to a business class passenger? One answer comes from the economist Frank Ramsey. Ramsey (1927) worked out that each user should cover the incremental costs (the peanuts) and that the fixed costs should be allocated in proportion to each user's price sensitivity—or the extent to which the quantity purchased varies with price. This rule is generally termed Ramsey pricing. Where such economies of scale exist, the Ramsey rule says that the least price sensitive group (business class) should pay the greatest proportion of the fixed costs through higher fares, and the most price sensitive group (the student traveler) should be charged the lowest proportion.

For water projects, users have wide variation in sensitivity to water prices—what economists refer to as the price elasticity of demand. Several studies have estimated the elasticity of demand for urban water users to be between −0.2 to −0.4 (i.e., when faced with a 100 percent increase in price, urban use would fall by 20% to 40%). Irrigated agriculture is more price responsive, with elasticities of demand for water ranging from −0.8 to −1.2 (implying a drop in use of 80 to 120 percent for a comparable 100 percent price increase). It follows that the practice, adopted by many water projects, of charging urban users higher prices than agricultural users can be justified as efficient, permitting the overall service area to benefit from scale economies. Similarly, urban water suppliers often charge commercial users more than residential users. In times of drought, those paying higher prices also are often provided with greater reliability (another economically efficient outcome). Such pricing principles are also common in the rail, electricity, and airline industries.[3] They would be appropriate for some of the Delta management alternatives examined in Chapters 7 and 8—

---

[3]Baumol and Willig (1981); Braeutigam (1979); Chessie System Railroads (1981); Damus (1981); Seneca (1973).

including the peripheral canal options and the near-Delta surface storage investments that might be used in the Opportunistic Delta scenario.

The key point is that if the beneficiary pays principle is to be implemented to cover all the costs of building a project, the sizing of the project must be balanced against different users' willingness to pay for different amounts of water. Project plans must also be backed by formal up-front financial commitments. Ramsey pricing is one way to balance these issues. It provides a standard method for efficiently allocating costs that users are willing to pay. Public statements about having users pay are not effective if the project design does not account for their observed willingness to pay.

Mitigating Environmental Damage

The above discussion focuses on apportioning the costs of new investments that directly benefit various stakeholders. It will also be appropriate to create programs of environmental mitigation for stakeholders who will benefit from whatever alternative is chosen, particularly when those benefits put pressures on environmental resources. These programs are already in place for water exporters, at least as a premise of existing Delta agreements. Exporters have been expected to participate financially in CALFED ecosystem restoration projects. Although agreement in this area has been slow, it is the basis of the Bay-Delta Conservation Plan now under development (see Chapter 5).

However, exporters are not the only group for whom environmental mitigation is relevant. As we saw in Chapter 6, in a typical year, upstream water users actually divert 80 percent more water from the Delta than exporters do. Although some upstream users have been involved in voluntary programs to contribute to the health of the Delta watersheds, there has been a tendency for both regulators and the environmental community to overlook upstream diversions and to focus exclusively on exporters.[4]

---

[4]Voluntary programs include the "Phase 8" agreements involving Sacramento River diverters and the Vernalis Adaptive Management Program (VAMP) within the San Joaquin River watershed. Under VAMP, some upstream users are being paid with public funds to alter the timing of their diversions to assist in maintaining adequate environmental stream flows.

Environmental mitigation should be required for the urbanization of Delta lands, given the irreversible changes caused by land development. One possibility would be set-aside requirements to maintain some lands for environmental uses. Such mitigations are already a standard practice for new development in many parts of the state. The Delta, with its unique environmental resources, should be no exception.

Environmental mitigation is also appropriate for ships using the Ports of Stockton and Sacramento, given the role of ballast water in introducing alien species. Present ballast water control requirements are too lenient to be of much value for the Delta. A ballast water fee could be imposed on shippers who do not undertake significant additional efforts. Tighter controls are also appropriate for horticultural, aquarium, bait, and other industries that deal with live organisms, all of which are likely sources of invasive species.

Public Sector Funding Roles

Even with application of the beneficiary pays principle to collective investments in water supply, flood control, and environmental mitigation, public funds will be needed to implement a more sustainable long-term solution for the Delta. State and federal taxpayer contributions are appropriate to help finance programs for which the general public is a beneficiary, such as environmental restoration. In some cases, these public benefits would include avoiding future public liabilities—a justification for taxpayer contributions to flood control and other emergency-preparedness measures. Public funds are also appropriate for programs considered important from the perspectives of equity and social justice—for instance, programs to provide safe drinking water to low-income rural communities. And finally, public funds can provide incentives to encourage various stakeholders to agree to actions that would generate overall social benefits that they might otherwise be reluctant to pursue. These last two reasons justify using bond monies or other public resources to finance programs to soften the costs of adjusting to Delta solutions.

Softening the Costs of Adjustment

No matter which Delta alternative is chosen, all users of Delta services will face some additional costs. In all cases, water exporters will need to make new investments to improve reliability and quality; under some alternatives, they would bear added water scarcity costs as well. Under any plan, some Delta farmers will go out of production because of island flooding; others will incur additional costs under regimes that feature fluctuating Delta salinity. Under most options, urban water users that pump directly from the Delta will need to alter their intake points, possibly building aqueducts to connect to reliable freshwater sources.

The increasing flood risks that accompany climate warming and sea level rise will also carry adjustment costs. Existing and planned urban areas behind Delta levees will need to invest in levee upgrades. The owners and users of the various types of infrastructure that crisscross the Delta will face additional costs for these same reasons. Suisun Marsh duck clubs will find it increasingly difficult to keep salt water from breaching their fragile levees and will eventually need to shut down or move elsewhere. And although recreational boating will continue in any likely future, alternatives that modify the channel network (e.g., Fortress Delta or the Armored-Island Aqueduct) could reduce revenues at some local harbors.

Candidates for Mitigation

Clearly, it is neither feasible nor desirable for state taxpayers to compensate all of these interests; doing business in the Delta is becoming more expensive because the current system is unsustainable, not because of the actions of the state or any one group. However, mitigation can soften the costs of adjustment for interests that will be particularly hard hit by changes to the status quo. For Delta management alternatives that move away from a freshwater Delta, this list includes Delta farmers and urban agencies that draw water directly from the Delta (notably, Contra Costa Water District). For alternatives that also significantly reduce water exports, this list includes farmers on the west side of the San Joaquin Valley and in the Tulare Basin. For alternatives that result in significant water transfers, this list might also include communities in the source regions. Other candidates could include owners of land that would benefit environmental goals—e.g., the Suisun Marsh duck clubs—or businesses that would be affected by changes in Delta channels.

There are no hard and fast rules for drawing up such a list. The goal of a mitigation process should be to encourage buy-in from interests that are likely to resist changes that could benefit the system as a whole. One consideration is legal standing. Under current agreements, Delta farmers and urban pumpers have protections on water quality (salinity) standards to the extent that these are affected by CVP and SWP exports. Another consideration is equity. As Chapters 6 and 8 showed, farmers in the Delta and in the San Joaquin Valley would lose out substantially under some alternatives. It makes sense to consider mitigation options to help ease transitions in these communities, whether or not there is a legal obligation to do so.

Mitigation Options

Mitigation does not imply a wholesale buy-out or coverage of all adjustment costs. Over time, the natural forces at work in the Delta will reduce the reliability of Delta services, requiring various groups to adjust anyway, largely at their own expense. Because almost all interests face worsening conditions, mitigations could be considered in relation to future "no action" conditions and effects, rather than in relation to some rosy, and unrealistic, continuation of current or past conditions.

Policies to soften adjustments could include a range of different forms of assistance. Many of these have been used in various contexts both in California and elsewhere.

Investment Cost Sharing

Cost sharing arrangements might be appropriate, for instance, if western Delta water users need to construct new pipelines or storage to allow the western Delta's salinity to fluctuate for ecosystem purposes. One example of a precedent for this kind of arrangement is the assistance provided to Los Angeles to reduce its diversions from the Mono Lake region. A state grant helped finance indoor conservation measures (notably, toilet retrofits) to reduce the city's water demand.

Financial Compensation

For farmland that will lose value, some form of financial compensation is likely to be appropriate. One option is outright land purchases. Precedents include Sherman Island, at the western edge of the Delta, which the state purchased in an attempt to ease water quality standards.[5] On the west side of the San Joaquin Valley, the federal government has already purchased some lands that have become unfarmable because of drainage problems, and there are proposals to make additional purchases as part of a settlement with farmers over the lack of drainage facilities.

Other options might be considered. In the Delta, where farming might continue indefinitely on some islands—depending on the patterns of island flooding and salinity intrusion—it may be beneficial to consider contracts under which farmers retain ownership but receive compensation for the eventual loss of farm income. Such a system has the advantage of letting farmers continue to manage the lands; they have more detailed knowledge and are likely to be better stewards than the state. Payments could be made up-front, as with a flood easement, or on withdrawal from farming. In either case, the land would not be eligible for development, and claims could not be made regarding future water quality standards.

One way to manage such a program would be through subsidized insurance or performance bonds. Currently, federal flood insurance programs offer protections to farm buildings, and federal crop insurance is available for flood damage to crops in any given year.[6] But there is currently no form of crop insurance that would cover the permanent loss of Delta farmland if islands become permanently flooded.

Although the Eco-Delta alternative is the only one that explicitly includes a transition to environmentally friendly farming in the Delta, such a transition would be appropriate under numerous alternatives for some Delta lands. Such farming practices would aim to restore soils, even sequestering carbon,[7] and would provide forage crops valuable for desirable terrestrial species, including sandhill cranes and Swainson's hawk. There

---

[5]This measure proved unsuccessful, because it was still necessary to maintain standards for industries operating in the western Delta. The state now leases these lands to farmers.

[6]Currently, farms in the Delta appear to have roughly the same rate of crop insurance coverage as farms in the rest of the state—with about 36 to 38 percent of all acreage insured. Crop insurance also covers damage from drought, hail, and other natural events.

[7]Carbon sequestration—or the capture of greenhouse gases—can be accomplished by growing certain perennial plants—like trees or tules—which store carbon captured from the atmosphere, particularly if these plants are then interred to prevent the carbon from reentering the atmosphere as the plants decay.

are clear precedents for this type of activity elsewhere in California and the nation. Conservation easements can be used to compensate farmers willing to make such transitions for the loss in other income. Within the Delta, it would be important to target lands that would generate the most environmental benefits.

All of the programs noted here have the goal of encouraging farmers to sign on to a program before island flooding occurs, to help advance the adoption of an overall solution package for the Delta. Another option is to compensate farmers for the loss of farmland once an island floods. To provide proper incentives, it would be appropriate to pay a higher price for land enrolled in the program early and to purchase flooded land at a discount.

Physical Substitution

For some activities, it may make sense to provide land rather than financial compensation. For instance, duck clubs currently in Suisun Marsh could be provided with lands farther east, making habitat in Suisun Marsh available for delta smelt and other threatened Delta species. Such physical solutions are a frequent feature of dam construction projects, which offer relocation possibilities for those displaced.

Community Mitigation Funds

In areas where a substantial proportion of lands discontinue agriculture, there may be economic consequences for the entire local community as well as for individual farmers. The compensation mechanisms described above can help farmers adjust to change, but they do not help the community—local farm laborers, agricultural input sellers and output processors, and even other local businesses and public services that may be affected by a loss of farm activity. Similar issues come up when there are large transfers of farm water outside of a region, as in the case of some recent transfers of Colorado River water to Southern California cities. In these cases, community mitigation funds have been set up to help various third parties to transition to new economic activities. Some parallels also exist in U.S. trade legislation, which provides adjustment assistance to workers and businesses displaced by imports. The early experiences in water transfer mitigation in California suggest the need for clear ground rules on eligibility and types of assistance (Hanak, 2003). As with water transfers, it is also important to recognize that such change does not always mean losses for a community, particularly when new opportunities arise in the local urban and recreational sectors.

Performance Bonds for Environmental Risk

For many of those depending on Delta services, uncertainties about the health of listed Delta species pose real costs, affecting the reliability of water supplies and land uses. This problem already exists, and it is likely to continue under any future Delta alternative. It may be possible to use performance bonds to capitalize some of these risks. Such bonds are used to cover the risks of cost overruns and delays in large construction projects. Performance bonds are essentially an insurance policy taken out based on the performance of a particular structure; if the structure does not perform to a set of specifications, the bond is paid to the owner. This arrangement provides assurance to the bond holder, who also has incentive (enforced by the insurer) to be prudent in constructing the structure.

Mitigations Versus Assurances

Mitigations and compensations differ from assurances. Assurances entail a guarantee of behavior or performance. The language of the CALFED era has been steeped in assurances, implying that such guarantees are possible. Mitigations and compensations do not assure specific future performance or actions. Rather, they provide a substitute for assured performance. Assurances of performance seem unreasonable for any likely Delta scenario, given the many uncertainties regarding the physical and biological dynamics of the Delta itself as well as long-term water availability in the Delta watersheds with climate change.

State and federal governments face an interesting policy problem regarding Delta mitigations. If consensus over a sustainable solution cannot be reached, nothing is likely to be done. The Delta is then likely to fail catastrophically, incurring major emergency expenses, plus restoration and remediation expenses—all under very unfavorable conditions. By investing in mitigations, some economically minor compensation costs (relative to California's $1.5 trillion per year economy) could be used to catalyze agreement on better long-term solutions for the Delta.

Governance Considerations

Although it is beyond the scope of this report to provide a detailed analysis of governance questions, we provide some thoughts on three central issues: improving coordination of land use in the Delta, managing environmental lands, and providing better incentives to manage water resources.

Bringing Delta Land Use into the Fold

As we have argued in the preceding chapters, long-term solutions for the Delta need to address land use as well as water and environmental goals. The CALFED process made important progress in coordinating agencies with responsibility for water and environmental management, but it overlooked land use agencies. Developing a governance framework that incorporates land use is particularly daunting in the Delta, given the current state of institutional fragmentation. Individual cities and counties are the permitting authorities for new development, and local reclamation districts are responsible for most decisions on levee maintenance and upgrades. There is little effective representation of larger regional and statewide interests in Delta land use decisions. This is a problem, given the broader public interest and considerable public investment in the Delta.

The Delta Protection Commission, established by the Delta Protection Act of 1992, is the only body representing regional interests in the Delta.[8] Its membership includes representatives from Delta cities, counties, and reclamation districts as well as various state agencies with Delta interests. Its primary purpose is to oversee land use and resource management issues in the Delta's primary zone, which the act reserved principally for agricultural, recreational, and environmental uses. Recently, the commission has begun serving as a regional forum for discussing growth issues more broadly. Although the commission may challenge land development that is inconsistent with the land use goals for the

---

[8]Delta cities and counties are members of three separate councils of government—the Association of Bay Area Governments, the Sacramento Council of Governments, and the San Joaquin County Council of Governments.

primary zone, it has no permitting authority and no ability to block land development.[9]

The state Reclamation Board currently does have the potential to exercise land use oversight in the Delta, through its authority to maintain the integrity of the flood control system. However, it has taken little interest in the Delta to date. Under current policies, it focuses only on those issues that either directly affect project levees (just over a third of all Delta levees—see Figure 2.2) or increase regional flood levels. As noted in Chapter 5, the board has come under criticism for its recent approval of the flood control plan for the River Islands housing development on Stewart Tract, with critics concerned that this decision did not adequately consider the implications for future flood risk either within the development itself or in neighboring areas.

Numerous other state and federal permitting agencies have the potential to affect land use in the Delta, including the California Department of Fish and Game and the U.S. Fish and Wildlife Service (species protection), the State Water Resources Control Board (water rights and water quality), the Army Corps of Engineers (flood control, navigation, and wetlands), and the Department of Water Resources and the U.S. Bureau of Reclamation (water contracts). However, none of these agencies have an institutional inclination for the regional management of resources of broad public interest.

The current lack of institutional authority for Delta land use, at a time when pressures to develop land resources are great, points to the need for a new approach. At a minimum, significant representation of state interests from outside the Delta is needed on the Reclamation Board and the Delta Protection Commission. More important, effective management of the Delta in the interest of the entire state will likely require an organization with more comprehensive oversight authority. Two models are the San

---

[9]For the second time in its 14-year history, the commission recently ordered a local authority to stop work while it reviews two appeals that challenge development in the primary zone. The case concerns a proposed 162-unit development in the northern Delta town of Clarksburg (unincorporated Yolo County). Litigation could result if there is disagreement between the county and the commission over the project's consistency with the provisions of the Delta Protection Act (Weiser, 2006c).

Francisco Bay Conservation and Development Commission (SFBCDC) and the California Coastal Commission.

These two bodies were created in response to pressures similar to those that now face the Delta. The SFBCDC, in operation since 1965, was established to tackle problems of uncoordinated development that were leading to the filling of the San Francisco Bay, which lost an average of four square miles per year between 1850 and 1960. The Coastal Commission, established in 1972, was created to ensure that land and water uses in the coastal zone are environmentally sustainable. Unlike the Delta Protection Commission, these two bodies have regulatory authority over a wide range of activities that have the potential to affect the beneficial uses of the bay and coastal resources. Both are authorized, under federal law, to exercise regulatory oversight of the actions of federal agencies. Both include a broad, representative membership.[10]

The SFBCDC's success has been truly remarkable—the San Francisco Bay is larger now than when the commission was created, and this has been achieved alongside the development of economic and recreational uses of the bay. As a regional entity, it provides a particularly interesting model for the Delta—which is part of the same valuable estuary as the San Francisco Bay.

An alternative management framework that has begun to draw interest in the Delta is the Habitat Conservation Plan/National Communities Conservation Plan model. As noted in Chapter 5, water exporters, state and federal fisheries agencies, and some environmental groups are in the process of developing such a framework, known as the Bay Delta Conservation Plan. In parts of Southern California, such plans have become very useful for making land use decisions that preserve open space and wildlife habitat at a regional scale. Designated areas are set aside for preserves, creating what some observers refer to as de facto urban limit lines. Both developers and environmental advocates see advantages in such an approach, which avoids piecemeal actions for habitat protection while providing more certainty to

---

[10]The SFBCDC has 27 members, including local land use authorities and various state and federal agencies. The Coastal Commission's 12 voting members include a mix of public members and local elected officials from various coastal areas; three state agencies have nonvoting status.

developers. These programs are also credited with facilitating fundraising for environmental mitigation.

Although the BDCP approach now under way may provide similar benefits to water users and Delta species, it is too limited in scope to serve as a comprehensive tool for governing Delta resources. In particular, it does not include local land use authorities. For this reason, we see the Bay Delta Conservation Plan process as a complement, rather than a substitute, for an institution like the SFBCDC for the Delta.

Managing Environmental Lands in the Delta

Another governance issue is the management of environmental lands. All Delta solutions will require more integrated management of water and land resources to foster improved habitat conditions for the Delta's aquatic and terrestrial species. As seen in Chapters 7 and 8, some Delta solutions have the potential to devote considerable resources to this goal. Environmental uses would, in many cases, be compatible with the further development of the Delta as a recreational destination. To manage these resources in a coordinated way, the establishment of various forms of public or nonprofit entities may be appropriate, including a state or national park or a nonprofit land trust. A land trust model is particularly compatible with the continued private management of some lands for eco-friendly agriculture. Land trusts across California and the country have played an important role in the development of conservation easement programs for farm and ranch lands. Some trusts play an active role in environmental land management as well.[11] One example within the Delta is the Cosumnes River Preserve, a 40,000 acre wildlife area managed by the Nature Conservancy and the U.S. Bureau of Land Management, in cooperation with other governmental and nonprofit partners.

Finance and Control of Water Facilities

In Chapter 8, we noted that stakeholder involvement will be important to develop good design and implementation rules for various Delta water management activities. New forms of stakeholder involvement are also likely to be an important part of any incentive package necessary to gain agreement on new water facility investments or the re-operation of

---

[11]For information on land trusts and a list of California organizations, see the website of the Land Trust Alliance (www.ltanet.org).

existing facilities. Finance and operational control of solutions are often unavoidably intertwined. Those who pay or invest have good reasons to want a role in the design, implementation, and operation of a solution. In the past, concerns over control of major new infrastructure facilities have led to unwillingness to either accept new facilities or pay for them.

One potential alternative is to assign shares of capacity of new (and perhaps existing) facilities to different parties with a stake in Delta water quality and water supply (upstream diverters, in-Delta users, exporters, and environmental agencies). Weekly or monthly pumping capacity would be allocated among the different agencies, with the share of a particular agency determined by its financial contribution or regulatory role. Under such a system, each party could affect the use of some infrastructure capacity to protect its interests, but there would be incentives for improved overall operations (e.g., through water exchanges and transfers). For instance, an environmental agency owning part of this capacity would have the option to limit diversions or to lease its share to other water users to generate revenues for environmental restoration activities in the Delta or upstream. Such an arrangement would give the environmental agency an incentive to allow pumping when it does little harm to fish, because it would provide revenues for other environmentally worthy activities.

The Environmental Water Account (EWA) established under CALFED is a prototype of this idea. It has sometimes been called a "water district for the environment," because it provides state and federal environmental managers with water resources that can be called on to regulate the operations of the CVP and SWP pumps to protect native fish species. However, this program has relied mostly on annual budget allocations rather than on a substantial permanent allotment of the water rights or project pumping capacity. As such, it has been subject to budgetary vagaries that may have limited its effectiveness (Rosekranz and Hayden, 2005).

Conclusions

Any viable long-term solution for the Delta's problems must encompass more than just a physical solution. It must also include fiscal and institutional solutions, requiring a political agreement. No Delta solution will be good for all parties. This was a delusion of the CALFED era, born of a now-depleted state and federal cash flow. Reaching a political agreement in the face of tradeoffs will be difficult and will likely require some compensations and mitigations. Such mitigations will require either greater external (state and federal) funding or increased payments from beneficiaries. In any event, beneficiaries will almost certainly need to pay most of the costs of fixing the Delta.

10. Conclusions and Recommendations

"The problem is not that there are problems. The problem is expecting otherwise and thinking that having problems is a problem."

Theodore Rubin

Conclusions

This report has five major conclusions:

1. The current management of the Delta is unsustainable for almost all stakeholders.
2. Recent improvement in the understanding of the Delta environment allows for more sustainable and innovative management.
3. Most users of Delta services have considerable ability to adapt economically to risk and change.
4. Several promising alternatives exist to current Delta management.
5. Significant political decisions will be needed to make major changes in the Delta.

We summarize each of these conclusions below and then offer some additional thoughts and recommendations.

Unsustainable Delta: Getting Worse Together

As we saw in Chapter 3, the Delta's future is unsustainable in its current form. Some key drivers of change in the Delta are largely beyond the control of stakeholders and policymakers. For example, climate warming is expected to contribute to sea level rise and to increased winter flows into the Delta, raising the likelihood of extreme flood events and levee failure. The increasing likelihood of a large earthquake affecting the Delta compounds this risk. Invasive species are posing increasing risks to the survival of key native species. Some invasive species, such as the Brazilian waterweed and the mitten crab, also pose growing risks to water supply. Furthermore, continued human population growth in California will raise the pressure on the Delta's land and water resources for recreation, housing, and water supply.

Other key drivers of change are more amenable to human intervention but only with major policy shifts. Land subsidence—compounding risk for levee failure—will continue with current farming practices. The accumulated effects of a century of land subsidence can only be reversed slowly. Urbanization in and around the Delta dramatically raises the potential damage from levee failure. It also poses a threat to the Delta's wildlife, by removing habitat. Contaminants from both agricultural and urban land use both within the Delta and in the upstream watersheds of the San Joaquin and Sacramento Valleys are major sources of water quality problems and concerns. New invasive species will continue to be a major threat to native species in the Delta, and current policies to prevent their arrival and limit their expansion are inadequate.

Any of these factors individually would cause great concern about the future of the Delta. In combination, they make the Delta's future look bleak. Given the potentially catastrophic nature of failure, we should prepare for a soft landing that allows us to accommodate and adapt to large-scale changes in the Delta, while allowing users of the Delta to extract themselves from their current untenable situation. The combined risk of Delta catastrophes for the state and for regions that depend on the Delta is too important to ignore any longer. Although crisis-response tools will be important, given the ever-present risk of levee failure, they are not a substitute for a new long-term solution. A sustained effort is needed to avoid such crises and their draining effects on the state's budget and economy. Without concerted action directed toward long-term solutions, all interests will be getting worse together.

Improved Understanding of the Delta Ecosystem

The common perception of the Delta as a stable freshwater habitat is wrong (see Chapters 2 and 4). The Delta is naturally a tidal system that historically has had salinities, water velocities, water clarity, and other characteristics that fluctuated widely across years, seasons, and tidal cycles, particularly in its western portions. Even today the volume of water moved daily by the tides far exceeds the amount of freshwater inflow, except under extremely wet conditions. This tidal influence is constantly moving salt water into the Delta. Thus, keeping the western Delta fresh during the dry summer months and in dry years requires greater reservoir releases to the Delta. In this way, saltwater intrusion is limited because most water in the Delta is confined to narrow leveed channels.

Given this artificial water regime, it is not surprising that the Delta's ecosystem is also highly altered and that many of its key native species are in decline, some to crisis levels. Restoring more natural fluctuations in salinity and other water quality and habitat conditions may be one of the most important ways to combat the many invasive species in the Delta. Many of these invaders are best adapted to stable freshwater or saltwater regimes, not to the fluctuating conditions to which many native species are adapted. A Delta that is heterogeneous and variable across space and time is more likely to support native species than is a homogeneously fresh or brackish Delta. Accepting the vision of a variable Delta, as opposed to the more commonly held vision of a static Delta, will allow for more sustainable and innovative management.

Economic Adaptations to a Changing Delta

Changes in the Delta will cause significant costs and some dislocations (see Chapters 6 and 8). However, most users of Delta services have considerable ability to adapt economically. As a result, these costs and dislocations need not be catastrophic for California's economy or society. However, these costs and dislocations will be much easier to handle if they are anticipated and dealt with in a systematic fashion, rather than in reaction to crises, such as levee failures.

For some, risk-mitigating investments and strategies can considerably diminish the costs of a catastrophic levee failure. One example is PG&E's strategy to increase redundancy of gas transmission lines with a new underground line in the Delta. Other examples include the plans of various water exporters to reduce dependency on the Delta by augmenting local sources and regional interties. For urbanizing Delta lands, strategies to increase flood protection may be able to reduce flood risk to acceptable levels (which need to be more realistically defined). Recreational users, such as duck hunters, are likely to have opportunities to relocate within the Delta as their current locations flood or otherwise change. In general, the Delta's role as a major recreation site will no doubt continue to expand, as the Northern California population grows. But the forms of recreational activity are likely to change and adapt with changes in the ecosystem and water management. Farming within the Delta is the economic interest with the least ability to adapt, because it relies on water and land uses that are unsustainable in many locations, even with substantial investments in the levee system. Even so, many farmers will be able to adjust to changes in water quality through alterations in their crop mix and irrigation practices. Public policies could help ease the transition away from Delta lands as they become economically unfarmable. Because they have nowhere else to go, the most vulnerable users of the Delta are the native species that rely on it for survival. Unlike human interests, their ability to adjust will depend entirely on society's stewardship decisions.

Alternative Management Strategies

Fortunately, the situation is not hopeless. There are promising alternative futures for the Delta (see Chapters 7 and 8). Some, like those based on the construction of a peripheral canal, have been proposed in the past. Others, like the Opportunistic Delta scenario involving only seasonal exports, are relatively new. No alternative will be ideal from all perspectives; some alternatives would preclude some current uses of the Delta entirely. Our analysis suggests that alternatives seeking to maintain the entire Delta as a freshwater system—along the lines of the current levee-centric policy—are incompatible with giving the Delta's native species a fighting chance to survive and prosper. The freshwater alternatives are also the least responsive to the drivers of change currently acting on the Delta. Various other alternatives would allow improvements in Delta habitat while permitting a variety of other beneficial uses.

The key to these alternatives is to use different parts of the Delta for different purposes. The most promising alternatives we discuss share similar strategies in this regard. Ecosystem restoration would be concentrated in the western Delta (where salinity would be allowed to fluctuate), Suisun Marsh, and the Delta's northwestern reaches, including the Cache Slough system and the Yolo Bypass. Agriculture would remain viable toward the north, east, and south; many of these areas could also contain urban development behind higher and stronger levees. These alternatives also provide the ability to continue water exports, either seasonally (Opportunistic or Eco-Delta) or year-round through one of several aqueduct alternatives (Armored-Island Aqueduct, Peripheral Canal Plus, or South Delta Restoration Aqueduct). All of these alternatives have different costs and risks, but each seems preferable to current conditions. Detailed knowledge, analysis, and discussion will be needed before identification of a "best" and politically viable alternative can be justified.

In each of these alternatives, some landowners and some water users would face particularly high adjustment costs, whereas others would benefit. Public policies would need to ensure that mitigation is available to distribute the costs equitably and reasonably (see Chapter 9). Mitigation could take the form of cost sharing for those whose adjustment costs are particularly high. This might be appropriate, for instance, if western Delta water users need to construct new pipelines or storage units to allow for ecosystem-based water quality fluctuations. Mitigation could also include policies to prevent further subsidence of agricultural lands or to buy out Delta farmers when their lands are no longer farmable because of flooding or water quality problems, an inevitable outcome in many of the Delta alternatives. With the resolution of major Delta policy issues, it would be easier to establish a more diversified, sustainable, and prosperous economy and ecosystem in the Delta.

Facing the Tradeoffs

A major change is needed in how Californians think about solutions to the Delta. The leitmotif of the approach adopted by CALFED was that "everyone would get better together," and it was assumed that this goal could be met by managing the Delta as a single unit, simultaneously achieving improvements in habitat, levees, water quality, and water supply reliability within the Delta and for exporters (Chapters 2 and 5). However, that approach was based on an insufficient appreciation of the risks of levee instability, an inadequate understanding of the importance of fluctuating conditions for some key native species, and the expectation of ample federal and state funding. Going forward, Californians will need to recognize that the Delta cannot be all things to all people. Tradeoffs are inevitable; the challenge will be to pursue an approach that yields the best outcomes overall, accompanied by strategies to reasonably compensate those who lose out. Incremental consensus-based solutions are unlikely to prevent a major ecological and economic disaster in the Delta.

Scientific and engineering studies and analyses can provide guidance on the types of alternatives that can meet the broadest range of goals. However, central to the decisions on a new course for the Delta will be the viability of funding mechanisms and governance institutions (see Chapter 9). Although CALFED fostered the beneficiary pays principle—whereby various economic interests were expected to contribute to program costs in proportion to the benefits they received—the default assumption, more often than not, was that the general public was the beneficiary. To wit, the proposed financing programs in 2000 and 2004 both relied heavily on funds from state and federal coffers. Although the assumption of federal largesse is now widely dismissed, many still look to the state to provide the bulk of the funding for Delta management. State general obligation bonds have funded most CALFED activities to date, and two bonds passed in November 2006 have allocated some $3 billion to $4 billion for flood control in the Central Valley and the Delta.

Yet the total initial and ongoing costs of any promising long-term Delta strategy will greatly exceed the availability of state bond funds, given other demands on public resources. (In the area of flood protection alone, great investments are needed to improve the protection of heavily urbanized areas upstream of the Delta, where the state has greater liability for flood damages.) For this reason, it will be essential to hammer out ground rules on funding contributions for both initial and ongoing operational expenses. The beneficiary pays principle will be especially relevant for any collective infrastructure investments that improve water supply reliability and reduce flood risk. The State Water Project was built on this principle. The financial contributions of water users and land development interests are likely to determine the most feasible investment choices. User finance of such investments is essential, given the other demands on public funds, such as ecosystem restoration. Under most scenarios, expenditures to purchase and manage lands for ecosystem restoration are likely to be considerable. Public funds will also be needed to contribute to mitigation solutions for those users who will lose out in whatever strategy is chosen. Creating long-term local dependency on state funding is undesirable from all perspectives, as it represents a great liability and drain on the state's coffers and provides an unreliable source of revenues beyond the control of local beneficiaries.

To best manage the tradeoffs in resource management within the Delta, there is a need for well-coordinated approaches that take into account not only water but also land use (see Chapter 9). The development pressures on the Delta are great, and the current institutional fragmentation in the Delta fosters piecemeal decisionmaking that will compound flood risks, irreversibly destroy valuable wildlife habitat, and impair water quality. Improved governance of Delta resources is necessary to protect the value of the Delta both for the region's residents and for the broader public interest.

Our analysis also suggests that the environmental community will need to consider new approaches to foster a healthy long-term future for the Delta ecosystem (see Chapter 5). The dominant assumption behind many recent environmental lawsuits—that the Delta's key problem is export volumes—may be only partially correct at best. If the various lawsuits now in play end up mandating reduced exports within the context of a static, freshwater Delta, the native species that policies are now aiming to protect are likely to suffer.

Recommendations

Our recommendations for the Delta fall into four categories:

1. Technical explorations of long-term solutions for the Delta are needed to inform the political process. Politics should not preempt the creative development, consideration, and comparative evaluation of alternatives.
2. Regional and statewide interests should be more forcefully represented in Delta land use decisions. These decisions have important implications for flood control, ecosystem health, and water supply and quality that extend well beyond the boundaries of Delta cities and counties. The Delta needs a strong regional permitting authority, along the lines of the San Francisco Bay Conservation and Development Commission or the California Coastal Commission.
3. To fund long-term investments in the Delta, the beneficiary pays principle needs to be resuscitated. Water users, urbanizing lands, and infrastructure users should all be expected to pay for investments from

which they benefit. Mitigation funds should be used to help ease the transition for those who will lose out from chosen alternatives.

4. Although it is premature to choose a long-term solution for the Delta without further technical investigation, Californians can take some steps now to move forward. To reduce the costs of a catastrophic levee failure in the Delta, investments in emergency preparedness are needed. To prepare the way for any long-term solution, discussions are also needed to implement some "no regrets" policies.

Technical Exploration of Solutions

1. **Create a technical track for developing Delta solutions.** For the most part, recent attempts to solve the Delta's problems have been politically driven. Under the rubric of "everyone getting better together," agencies and other stakeholders sought to negotiate solutions based on what was politically acceptable. Despite considerable investments of time and money, this approach has not resulted in an acceptable or workable solution. Now we are all getting worse together. This failure has led to calls for solutions, largely derived from past proposals, which maintain the Delta in its present configuration. Despite improvements in our understanding of the Delta ecosystem and the economy of California, little in the way of new solutions has been developed or proposed. The political track of any Delta solution is important and necessary, but it can be better informed and seeded with more viable answers by a technical track that would develop and explore new ideas and adapt older solutions to current conditions.

2. **Establish an institutional framework to support the development of solutions to the Delta's problems and to bring scientifically and economically promising alternatives to the attention of political authorities.** This activity needs to take a long-term view and avoid crisis-driven responses to short-term political thinking. It should have some political independence, an appropriately sized budget, the technical capability to creatively and competently explore and eliminate alternatives, and the management capability to direct multidisciplinary research and development. CALFED was supposed to have these abilities, but its direction, funds, and energy became dissipated in politics and the effort to please all stakeholders. At the turn of the

last century, California's Debris Commission had a similar problem-solving role (see Chapter 2). In taking a long view, it paved the way for fundamentally different and more successful flood management in the Central Valley, leading to the introduction of flood bypasses. The current technical efforts examining the pelagic organism decline and the risks to Delta levees focus rather narrowly on specific aspects of the Delta's problems, and the current policy efforts—including the Delta Vision process—currently lack a substantial technical component. Technical and policy endeavors need some independence within a larger framework.

3. **Launch a problem-solving research and development program.** The science effort regarding the Delta is in need of an overhaul. The Delta is a multidisciplinary problem, not a single-focus research topic. Much past research on the Delta and its problems has been associated with agency data collection or basic agency, academic, and disciplinary research. Although such efforts have helped improve our understanding of the Delta, they have not provided an efficient or effective process to support decisionmaking. A directed problem-solving research and development program aimed primarily at developing and informing the analysis of promising solutions is needed (see Chapters 4 and 8). This program would include some continued basic research, but most effort would be directed toward developing and evaluating solutions. Ecosystem adaptive management experiments (supported by quantification and computer modeling), levee replacement, island land management, flood control, and integrative system design activities should receive greater attention in a problem-solving framework.

4. **Consider the Delta's water delivery problems in a broad context.** The foremost physical problem in the Delta needing a physical solution is delivery of fresh water through or around the Delta because this water is a key factor driving California's economic engine. And some promising solutions exist. Potential options extend beyond a peripheral canal. Our work suggests that an armored-island aqueduct, a south Delta restoration canal, opportunistic pumping, and perhaps even an environmentally reoriented Delta management scheme all show promise and merit further exploration (see Chapter 8). Any physical

solution for water delivery must be accomplished in the broader context of developing a more sustainable Delta environment.

5. **Eliminate some solutions to the Delta's water delivery problems from further consideration.** To reduce investments in scarce time, expertise, and resources in evaluating Delta alternatives, some potential Delta options are not worth further exploration (see Chapter 8). These include the traditional levee-centric approach, the building of downstream physical barriers to seawater, the large expansion of on-stream surface water storage, and the idea of ending all export pumping. These are physically unreasonable solutions to the Delta's water delivery problems, and they perform so poorly in economic and environmental terms as to be nonviable.

6. **Approach the Delta as a diverse and variable system rather than as a monolith.** A diversified and variable Delta by design is likely to perform better than the freshwater Delta that has been artificially maintained over the last 60 years. Better solutions are likely to emerge if the Delta is not treated homogeneously (see Chapter 4). Historically, the Delta naturally contained diverse habitats that varied across years, seasons, and tidal cycles in terms of salinity, water residence time, turbidity, water velocity, elevation, and other physical habitat conditions. Reintroducing and extending this diversity, by specializing parts of the Delta for wildlife habitat, agriculture, urban, recreation, water supply, and other human purposes, seems promising.

Governing and Financing Change

1. **Create stronger regional and statewide representation in Delta land use decisions.** Local land use interests in the Delta are well represented by local cities, counties, water agencies, and reclamation districts, but these institutions are fragmented. There is little effective representation of larger regional and statewide interests in Delta land use decisions (see Chapter 9). An institutional disconnect exists between local land use planning and the broader public interest—and considerable public investments—in the Delta.

No current agencies or institutions have broad authority to oversee land use decisions in the Delta. The existing Delta Protection Commission, whose role is to foster continued agricultural, recreational, and environmental uses of most Delta lowlands, is a weak institution without permitting authority. To date, the State Reclamation Board has taken little interest in the Delta and, under current policies, focuses only on those issues that either directly affect federally authorized project levees or increase regional flood stage. The CALFED Bay-Delta Authority has no direct influence over land use decisions. State and federal permitting agencies, including DWR, the Department of Fish and Game, SWRCB, the U.S. Bureau of Reclamation, the U.S. Fish and Wildlife Service, and the Army Corps of Engineers, have no institutional inclination for regional management of resources of broad public interest.

The "all politics are local" adage applies well to the Delta, yet local land use decisions there affect the entire state. A new approach is needed that, at minimum, provides for *significant* representation of state interests from outside the Delta on decisionmaking bodies (such as the State Reclamation Board or the Delta Protection Commission). Effective management of the Delta in the interest of the entire state will require an organization modeled after the California Coastal Commission or the San Francisco Bay Conservation and Development Commission.

2. **Give direct beneficiaries primary responsibility for paying for Delta solutions** (see Chapter 9). Urban development should pay directly for its own flood protection (including both capital and maintenance costs) with protection set at appropriately high levels (exceeding 200-year average recurrence for concentrated development). It should also contribute substantially to environmental offsets, given the significant, irreversible changes it causes. Direct and indirect exporters of water from the Delta should pay for infrastructure that directly benefits them and should contribute to ecosystem restoration necessary to offset the effects of water exports. Other Delta infrastructure providers (roads, pipelines, power lines, etc.) should be expected to pay for their own facilities. A ballast water fee or tax should apply to shippers who do not undertake significant efforts to preclude the introduction of invasive alien species, and tighter controls should be imposed on horticultural, aquarium, bait, and other industries that deal with live organisms. It should be acknowledged that agricultural activities, though principal

beneficiaries of many proposed Delta improvements, will not be able to raise large quantities of funding to address most Delta problems.

Public funds, such as those raised through general obligation bonds, should be reserved for the truly public components of the investment program, such as ecosystem restoration and mitigation for those who lose out as Delta strategies shift. Public funds can also complement private funds for some investments that have both private and public goods characteristics, such as some flood control or environmental water supplies. Failure to develop an effective funding mechanism is likely to lead to financial catastrophes for state and local interests in the future, in the wake of natural catastrophes.

Funding and control of water export facilities and operations are likely to be intertwined. In the past, concerns over control of major new infrastructure facilities have led to unwillingness to either accept new facilities or pay for them. One potential alternative is to assign shares of capacity of new (and perhaps existing) facilities to different parties with a stake in Delta water quality and water supply (upstream diverters, in-Delta users, exporters, and environmental agencies) (see Chapter 9). Under such a system, each party could affect the use of some infrastructure capacity to protect its interests, but there would be incentives for improved overall operations (e.g., through water exchanges and transfers).

3. **Establish mitigation and compensation mechanisms to support the implementation of any alternative.** Not everyone will get what they want or what they have been used to getting from the Delta. In some cases, providing money or alternative land might compensate for changing or eliminating uses of water or land that hinder broader progress (see Chapter 9).

Urgent Items for Policy Debate and Action

1. **Make essential emergency preparedness investments.** This report has focused on long-term solutions for the Delta, which will take some time to put into place. In the short term, it is crucial to take steps to mitigate the costs of a sudden failure of Delta levees (see Chapter 8). For all agencies relying on Delta waters, this means developing plans to ride out an extended export outage. With measures such as regional

interties, water sharing agreements, local supply development and drought contingency plans, the costs of losing a year of Delta exports can be reduced by a factor of 10 (see Chapter 6). Other infrastructure providers that rely on the Delta, such as Caltrans, the railroads, and power companies, need similar contingency plans, and should consider making new investments in their networks to make them less susceptible to levee failure. PG&E's investment in a buried gas pipeline is a case in point. The continuation of a program for the rapid repair of critical levees—such as the one launched in 2006—and the development of emergency flood response plans are also key.

2. **Implement a "no regrets" strategy for the Delta** (see Chapter 8). First, given the great urbanization pressures on the Delta, several actions are needed now to avoid irreversible consequences. These include establishing an improved regional governance structure, instituting a program to set aside or purchase key habitat, and creating adequate, coherent flood control guidelines for urbanizing lands.

Second, because not all Delta islands have the same strategic value, in terms of either economic assets (including homes and infrastructure) or water supply, policymakers should develop a "do not resuscitate" list in the event of levee failure. Making such decisions now could avoid costly expenditures on islands that are of low strategic value, while creating opportunities to experiment with a more variable Delta environment. This list could be coupled with insurance or buy-out programs for lost farmlands on these islands.

Third, a substantive improvement in the Delta ecosystem, germane for any long-term Delta solution, could be made with habitat restoration actions in the Suisun Marsh and Cache Slough regions. A variety of other "no regrets" actions were started under CALFED, including groundwater banking, water use efficiency, water marketing, and environmental water account activities (see Chapter 2). These actions should be continued, albeit with support predominantly from water users.

Forging a New Path Forward

The Delta's many problems have sparked a crisis of confidence on the part of its many stakeholders. The CALFED process, which has been responsible for crafting solutions in the Delta since the mid-1990s, is now widely perceived as having failed to meet its objectives. That process was forged under the urgent threat of a regulatory hammer—a severe cut-back in pumping to meet federal water quality standards for the Delta (see Chapter 2). CALFED's failure lay in the course chosen for crafting solutions. Achieving political consensus was favored over making tough choices among alternatives, and it was assumed that taxpayer largesse would foot any bill. In the past, major innovations in Delta management have required dire external pressure—real or threatened—from droughts, floods, lawsuits, or federal or state government. The question going forward is whether today's crisis in the Delta can spur stakeholders and the state to action with a new strategy that acknowledges the fact that some will gain and some will lose out as the Delta changes. The future of this unique ecosystem and regional land resource and of the state's water supply system depends on the answer. All Californians are likely to see benefits (and costs) from a comprehensive long-term solution. Otherwise, we will all see only costs.

Appendix A

Paradigm Shifts in Our Understanding of the San Francisco Estuary as an Ecosystem[1]

"In all affairs it's a healthy thing now and then to hang a question mark on the things you have long taken for granted."

Bertrand Russell

The San Francisco Estuary has a long history of being important to Euro-American endeavors in California. In the 19th century, it supported commercial fisheries and was a major transportation corridor, while the Delta and Suisun Marsh gradually became developed as farmland (and then as freshwater marsh managed for waterfowl). These functions continued well into the 20th century, while urban areas expanded, filling in marshlands and dumping large amounts of raw sewage into the water. The basic attitude of this era was that the natural environments would take care of themselves and their health was subservient to human needs. When the State Water Project was built in the 1960s, some restrictions were included to protect Suisun Marsh and the Delta, recognizing that freshwater outflows were needed to protect duck hunting, agriculture, and western Delta cities as well as to feed water to the pumps in the southern Delta. The passage of the Clean Water Act in 1972 resulted in the rapid cleanup of sewage treatment plants around the estuary. This and other state and federal laws passed in the 1970s reflected a changing public attitude toward the need for a healthy environment, especially to protect human health. These changes in attitude and ways of managing the San Francisco

---

[1] Peter Moyle is largely responsible for the material in this appendix.

Estuary reflect paradigm shifts in our understanding of how ecosystems work, including the human role in them.[2]

The first major paradigm shift was from the concept that ecosystems were infinitely resilient and existed for humans to use as they pleased, with no harmful consequences resulting from such use. The shift was toward the view that ecosystems could be greatly harmed by human activity, often to our own detriment, but that changes were reversible. This led to the concept that ecosystems could be restored to their former states. Ecological theory, developing rapidly in the latter half of the 20th century, originally supported the restoration concept. The paradigm was stated succinctly as the "balance of nature": An ecosystem knocked off center would return to its ideal, desirable state if allowed to do so. By the 1990s, however, this paradigm had shifted to the paradigm that "the only constant is change," that ecosystems are constantly changing in response to multiple factors, especially rapid and long-term shifts in climate and geology. Human activity by and large accelerates natural change and forces it in directions that are often undesirable from the perspective of native organisms and, increasingly, humans themselves. These changes are often irreversible. In a situation such as the Delta, "restoration" means choosing the attributes and organisms regarded as desirable and then finding ways to manage the system for desired conditions. Rosenzweig (2004) prefers to call such actions "reconciliation" rather than restoration because the managed system is going to remain human-dominated no matter what.

Not surprisingly, shifts in societal perceptions of the environment and in ecological understanding are reflected in actions taken to manage the Delta's estuarine ecosystem, although the target of management has usually been aquatic organisms, especially fish. The motivation for management has been declines in important fish species, initially those that supported fisheries (e.g., striped bass, Chinook salmon, sturgeon) but more recently native species perceived as being at risk of extinction (e.g., delta smelt, splittail, winter-run Chinook salmon). These declines have been under way for a long time. Arguably, the rate and extent of declines could have been

---

[2]A paradigm is a "set of interrelated assumptions on the functioning of a system that form a conceptual framework" (Craine, 2006, p. 449). When the assumptions change as the result of new information, a shift to a new paradigm or understanding can occur.

reduced if the biologists advising managers had had a better understanding of the Delta ecosystem.

Indeed, many of the basic concepts of how the system worked—which formed the basis for decisions regulating outflow by the State Water Resources Control Board—were wrong or inadequate. The misconceptions start with calling the upper estuary the Delta, implying that it was created primarily by deposits of river sediment, as are other deltas. Instead, it was created as a unique marsh/peat system where slowly rising sea level in a low-lying area created the anoxic conditions suitable for the deposition of organic material from marsh plants, supplemented by deposits of river sediments. This initial misconception helped to fix the idea of the Delta as the upper portion of a more or less linear, river-driven estuary, such as those found in the eastern United States.[3] Thanks to research conducted over the last 20 years, our understanding of how the Delta and estuary work has improved greatly, resulting in the paradigm shifts discussed here.

Listed below are major paradigm shifts that have taken place or are starting to take place regarding the San Francisco Estuary, especially the Delta, along with shifts in some key underlying assumptions that support the paradigms. We have tried to state succinctly the new paradigm or assumption and then the one (old) that it has replaced.

Uniqueness of the San Francisco Estuary

- **New paradigm:** The San Francisco Estuary is unique in many attributes, especially its complex tidal hydrodynamics and hydrology. **Old paradigm:** The San Francisco Estuary works on the simple predictable model of East Coast estuaries with linear gradients of temperature and salinity controlled by outflow with edging marshes, both salt and fresh water, supporting biotic productivity and diversity
  - **New assumption:** Daily tidal excursions have more hydrodynamic influence on the ecology of the estuary than outflows do, especially in the western and central Delta, except during high outflow

---

[3]During the period of Delta formation, the accumulation of organic matter made it a net sink for carbon; carbon dioxide is a major greenhouse gas contributing to global warming. Since the advent of agriculture, the carbon historically locked up in Delta peat has been released into the atmosphere. Stopping or reversing this process could contribute to slowing climate warming.

events. **Old assumption:** The most important hydrodynamic force in the ecology of estuary is freshwater outflows, especially within the Delta.

- New assumption: Striped bass are only one part of the estuary ecosystem and conditions that benefit them do not necessarily benefit native organisms. **Old assumption:** If the estuary is managed for striped bass (an East Coast species), all other organisms, but especially other fish, will benefit.
- New assumption: Creating more shallow freshwater habitat benefits mainly alien species in the Delta. Development of dendritic channel patterns with residence time diversity might be a key to restoration. **Old assumption:** Creating more shallow freshwater habitat is the key to making the Delta more friendly to native species.

Invasive Species

- **New paradigm:** Alien species are a major and growing problem that significantly inhibits our ability to manage for desirable species. **Old paradigm:** Alien (nonnative) species are a minor problem or provide more benefits than problems.
  - New assumption: Some alien species have major effects on ecosystem structure and function, with negative effects on highly valued species. **Old assumption:** Alien species mainly increase biotic diversity and harm mainly low-value native species.

Interdependence

- **New paradigm:** Changes in the management of one part of the entire estuary system affect other parts. **Old paradigm:** The major parts of San Francisco Estuary can be managed independently.
  - New assumption: All areas are part of the estuary and can change states in response to outflow and climatic conditions. **Old assumption:** The Delta is a freshwater system, Suisun Bay and Marsh are brackish water systems, and San Francisco Bay is a marine system.

- **New assumption:** Floodplains are of major ecological importance for many organisms, including salmon and other native fish as well as migratory birds, and they affect estuarine function. **Old assumption:** Floodplains such as the Yolo Bypass have little ecological importance and are independent of the estuary.
- **New assumption:** Suisun Marsh is an integral part of the estuary ecosystem and its future is closely tied to that of the Delta. **Old assumption:** Suisun Marsh is independent of the rest of the estuary.

Stability

- **New paradigm:** Delta landscapes will undergo dramatic changes as the result of natural and human-caused forces such as sea level rise, flooding, climate, and subsidence.
  Old paradigm: The Delta is a stable geographic entity in its present configuration.
  - **New assumption:** The Delta will most likely change dramatically in the next 50 years. **Old assumption:** The Delta can be maintained pretty much in its present configuration indefinitely.
  - **New assumption:** There will still be an ecosystem if the configuration of the Delta changes; some changes may actually be an improvement (from a fish perspective) over the existing ecosystem. **Old assumption:** A change in Delta configuration will destroy the present ecosystem.
  - **New assumption:** Management of the Delta requires a flexible, adaptive approach, where objectives change in response to improved knowledge of the system. **Old assumption:** Management of the Delta requires fixed, achievable objectives.
  - **New assumption:** All Delta levees will or can fail; building bigger levees just reduces the frequency of failure. **Old assumption:** Levees can be built in the Delta that will not fail.
  - **New assumption:** Agriculture is an unsustainable use of land and water in many parts of the Delta, which may instead be best suited for recreation or natural habitats. **Old assumption:** The best and most desirable use of land and water in the Delta is agriculture.

Delta Pumping

- **New paradigm:** The big pumps in the southern Delta are one of several causes of fish declines and their effect depends on species, export volume, and timing of water diversions.
  Old paradigm: The big SWP and CVP pumps in the southern Delta are the biggest cause of fish declines in the estuary.
  - **New assumption:** Entrainment of fish at the power plants at Pittsburg and Antioch is potentially a major source of mortality, especially of larval fish, that could significantly contribute to the pelagic organism decline. **Old assumption:** Entrainment of fish in the power plants at Pittsburg and Antioch is a minor source of fish mortality and can be ignored.
  - **New assumption:** Changes in ocean conditions have major effects on the Delta by affecting rainfall and other aspects of climate, as well as the survival rates of anadromous fish such as Chinook salmon. **Old assumption:** Changes in ocean conditions (e.g., El Niño events, Pacific Decadal Oscillation) have no effect on the Delta.
  - **New assumption:** Hatcheries are an important contributor to the decline of wild salmon and steelhead populations and confuse salmonid restoration work in the Delta because of our inability to determine the effects on hatchery versus wild fish. **Old assumption:** Hatcheries have no effect on wild populations of salmon and steelhead.
  - **New assumption:** Although chronic toxicants continue to be a problem, episodic toxic events (e.g., from storm drains and agricultural applications) are also a major problem (e.g., they can alter food webs). **Old assumption:** Chronic toxicants (e.g., heavy metals, persistent pesticides) are the major problems with toxic compounds in the estuary.

Appendix B
Stakeholder Consultations

The following people generously agreed to discuss Delta issues with us during the course of this research through in-person meetings or telephone conversations. Participants in two technical workshops are listed in the Preface and Acknowledgments section.

Kome Ajise, Caltrans
Chuck Armor, Department of Fish and Game
Gary Bobker, Bay Institute
Alf Brandt, Assembly Water, Parks, and Wildlife Committee
John Cain, Natural Heritage Institute
Pam Carder, City of Lathrop
Jeff Carroll, PG&E
Tom Clark, Kern County Water Agency
Marci Coglianese, former mayor of Rio Vista
Gil Cosio, MBK Engineering
Martha Davis, Inland Empire Utilities Agency
Susan Dell'Osso, Cambay Group
Tom Erb, Los Angeles Department of Water and Power
Linda Fiack, Delta Protection Commission
Jamie Fordyce, Environmental Defense
Tony Francois, California Farm Bureau
Dave Fullerton, Metropolitan Water District of Southern California
Greg Gartrell, Contra Costa Water District
Tom Graff, Environmental Defense
Dorothy Green, California Water Impact Network
Joseph Grindstaff, CALFED Bay Delta Program
David Guy, Northern California Water Association
Les Harder, Department of Water Resources
Ann Hayden, Environmental Defense
Bruce Herbold, U.S. Environmental Protection Agency
Alex Hildebrand, South Delta Water Agency

Mary Hildebrand, South Delta Water Agency
Doug Holland, Assembly Republican Caucus
Jerry Johns, Department of Water Resources
Randy Kanouse, East Bay Municipal Utilities District
Gregg Lemker, PG&E
Steve MacCauley, California Urban Water Agencies
Senator Michael Machado
Steve McCarthy, Senate Republican Caucus
Rod Meade, CALFED Bay Delta Program
Gerry Meral
B. J. Miller, consultant
Laura King Moon, State Water Contractors
Anson Moran, Delta Wetlands Project
Phil Nails, Assembly Republican Caucus
Barry Nelson, Natural Resources Defense Fund
Dan Nelson, San Luis and Delta Mendota Water Authority
Chris Neudeck, Kjeldsen, Sinnock and Neudeck, Inc.
Mary Nichols, Institute for the Environment, UCLA
Dennis O'Connor, Senate Natural Resources Committee
Tim Quinn, Metropolitan Water District of Southern California
Spreck Rosekranz, Environmental Defense
Frances Spivy Weber, Mono Lake Commission
Mike Wade, Farm Water Coalition
Walt Wadlow, Santa Clara Valley Water District
Brent Walthall, Kern County Water Agency
Bethany Westfall, Senator Machado's Office
Rebecca Willis, City of Oakley
Tom Zuckerman, Central Delta Water Agency

Appendix C
CALVIN Model and Results

"What would life be without arithmetic, but a scene of horrors."

Reverend Sydney Smith (1935)

CALVIN is a computer model developed to explore how California's water supply system would function under very different conditions and policies. It is an integrated economic-engineering optimization model, meaning that it incorporates many engineering aspects of the water supply system (infrastructure and hydrology) along with economic management purposes and environmental constraints on the system's operations. As an optimization model, it operates to maximize net statewide economic benefits of urban and agricultural water supply, within hydrologic, infrastructure, environmental, and other introduced policy constraints. The model was developed as a strategic screening model to identify promising solutions and to provide preliminary estimates of some major economic benefits and costs for California's complex statewide water supply system (Figures C.1 and C.2). CALVIN has been applied to explore the economic value and operational implications of new water facilities (Jenkins et al., 2001; Draper et al., 2003, Jenkins et al., 2004) and has been used in various other applications, including climate change (Tanaka et al., 2006; Medellin et al., 2006; Lund et al., 2003), conjunctive use (Pulido-Velázquez, Jenkins, and Lund, 2004), dam removal (Null and Lund, 2006), water marketing (Newlin et al., 2002; Jenkins et al., 2004), among others (Van Lienden and Lund, 2004). These sources and the CALVIN web site provide full explanations of the model and its major limitations (http://cee.engr.ucdavis.edu/faculty/lund/CALVIN/). The CALVIN model employs the U.S. Army Corps of Engineers' HEC-PRM reservoir operation optimization software.

Model Assumptions and Coverage

CALVIN includes California's entire intertied water supply system, including all major areas that depend directly or indirectly on Delta flows.

The model includes alternative surface and groundwater supplies, economic representations of operating costs, and the economic costs of water scarcity (or shortage) to urban and agricultural water users (implicitly including the costs of water conservation responses for urban and agricultural water users). Many of the water management options included in the model are listed in Table 6.2.

The version of the model employed here estimates water demands for the year 2050 for a statewide population of 65 million (Medellin et al., 2006), using projections developed in 2002 at UC Berkeley (Landis and Reilly, 2002). A set of 72-year monthly statewide inflows has been used to represent the historical variability of wet and dry years and seasons typically seen in California. Figure C.1 shows the extent of water demands and infrastructure modeled in CALVIN. All major surface and groundwater supply and conveyance facilities in California's intertied water system are included. All major urban and agricultural water demands also are included and represented economically.

We consider three scenarios: a base case, a no-exports case, and an increased-outflow case. The base case for 2050 assumes that water agencies will complete currently planned infrastructure enhancements and undertake additional investments in both supply-side and demand-side alternatives to meet 2050 demands cost-effectively. The no-exports scenario assumes that some additional intertie capacity is constructed, mostly where some aqueducts currently cross or are nearby, and it allows water users to make additional cost-effective investments in water supplies and demand management. Similar assumptions hold for the scenarios involving increased minimum Delta outflows. Because the no-exports alternative reduces opportunities for water transfers (preventing Sacramento Valley and east side San Joaquin Valley water users from moving water through the Delta), the patterns of supply investments, water marketing, and operational opportunities are different from those in the other two scenarios. To see this, we compare the three scenarios (base case, no exports, and increased outflows); in the increased-outflow scenario, total water use is cut by the same amount (5 maf) as it is under the no-exports case.

For the no-exports and increased-outflow model runs, some additional intertie conveyance capacity was added, reflecting projects that are planned or currently under way. These new interties include capacity to divert water from the Sacramento River at Freeport to the Mokelumne River Aqueduct,

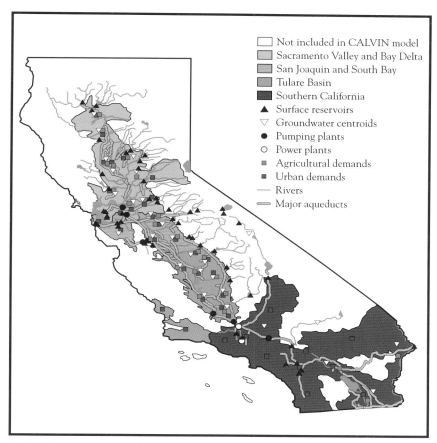

Not included in CALVIN model
Sacramento Valley and Bay Delta
San Joaquin and South Bay
Tulare Basin
Southern California
▲ Surface reservoirs
▽ Groundwater centroids
● Pumping plants
○ Power plants
■ Agricultural demands
■ Urban demands
— Rivers
▭ Major aqueducts

NOTE: Groundwater centroids refer to the center of active groundwater basins.

Figure C.1—CALVIN Demand Areas and Major Infrastructure and Inflows

a diversion from the Mokelumne River Aqueduct to the Contra Costa
Canal, and an intertie between Hayward and EBMUD. These interties
allow the Contra Costa Water District, currently served exclusively by Delta
pumping, access to alternative supplies, and they provide the Santa Clara
Valley, San Francisco, EBMUD, and others with additional water purchase,
sale, and management opportunities.

In addition, in the no-exports and increased-outflow model runs,
urban coastal areas were assumed to have access to desalinated seawater at a
cost of $1,400 per acre-foot and all urban areas were assumed to have access
to reused wastewater up to 50 percent of their allowable wastewater flows at

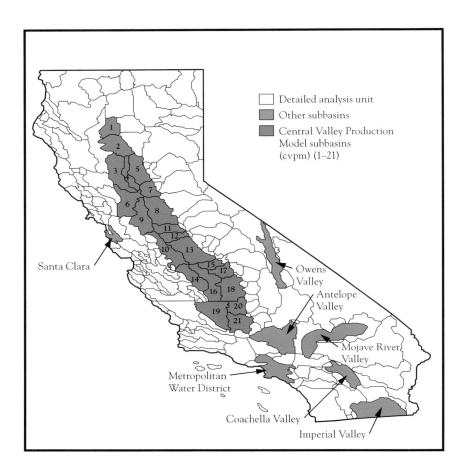

Figure C.2—Agricultural Regions in CALVIN

a cost of $1,000 per acre-foot. Household and industrial water conservation is also assumed to be available at a variable cost represented by a constant elasticity of demand curve for residential users and survey-based cost functions for industrial users (Jenkins, Lund, and Howitt, 2003). In the model and in reality, traditional water supplies from surface and ground waters incur operating costs for pumping, recharge, and water treatment, and some relatively saline urban supplies also incur costs to customers because of their salinity (Jenkins et al., 2001).

The No-Exports Scenario

We are unlikely to know the exact costs and operations undertaken if direct water exports from the Delta were abandoned. CALVIN model results under this set of conditions suggest operations and water management that would minimize the overall costs of such conditions. These assume considerable preparations in terms of interties and operating agreements among agencies, with supplemental water market and exchange agreements.

In the no-exports alternative, overall agriculture scarcity increased by 4,723 taf per year and urban scarcity increased by 288 taf per year. All of the scarcity increase occurred south of the Delta, with the increase in agricultural scarcity limited to the western San Joaquin Valley and Tulare Basin and the majority of the urban scarcity increase occurring in Southern California (Table C.1).

Agricultural users south of the Delta bear the brunt of the water supply cuts and economic costs associated with a no-exports alternative. However, as illustrated in Figures C.3 and C.4, these effects are uneven. Agricultural

Table C.1

Sectoral Water Scarcity, by Region

	Annual Average Scarcity (taf/year)		Scarcity as a % of Demand	
	Delta Exports	No Exports	Delta Exports	No Exports
Agriculture				
Sacramento Valley	318	137	3	1
San Joaquin Valley and Tulare Basin	1,632	6,535	10	39
Southern California	941	941	29	29
Statewide	*2,891*	*7,614*	*10*	*26*
Urban				
Sacramento Valley	0	0	0	0
San Joaquin Valley and Tulare Basin	0	29	0	1
Southern California	60	318	1	4
Statewide	*60*	*347*	*0*	*3*

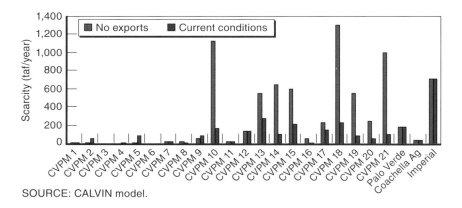

SOURCE: CALVIN model.

Figure C.3—Annual Average Agricultural Water Scarcity by Agricultural Area for 2050 Conditions with No Direct Delta Exports

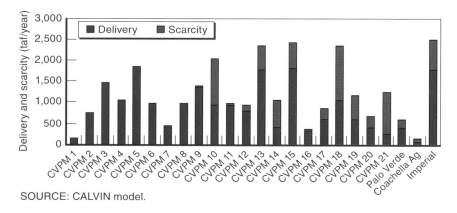

SOURCE: CALVIN model.

Figure C.4—Annual Average Agricultural Water Deliveries and Scarcity by Agricultural Area for 2050 Conditions with No Direct Delta Exports

areas depending directly on streams flowing from the Sierra Nevada Mountains (primarily on the east side of the San Joaquin Valley, Central Valley Production Model (CVPM) regions 11, 12, 16, and 17, located in Figure C.2) are much less affected by eliminating Delta exports, as their water supplies do not depend on the Delta and they cannot connect to other agricultural regions farther south and west without going through the Delta. Water districts that depend more on Delta pumping—on the west side of the San Joaquin Valley and in the Tulare Basin (CVPM regions 10, 14, 19, and 21)—are most severely affected. Agricultural areas dependent on San Joaquin River diversions at Friant Dam and Tulare Basin inflows also are affected, because these remain the only transportable surface waters that can serve Tulare Basin (CVPM 18, 19, 20, and 21), southern San Joaquin Basin agricultural users (CVPM 13), and urban Southern California. However, many farmers with rights to Friant-Kern and local Tulare surface waters are likely to do well financially through sales of this scarce water to cities in Southern California. Despite the ending of Delta exports, Metropolitan Water District of Southern California maintains reduced deliveries of water from the California Aqueduct, averaging 1.3 maf per year (0.9 maf per year less than when Delta exports are allowed); these supplies are purchased from Tulare Basin inflows and San Joaquin River diversions at Friant Dam.

Average urban water delivery and scarcity volumes appear in Figure C.5. Southern California cities from Ventura to San Diego, having lost 2,248 taf per year in supplies from the Delta, purchase 1,321 taf per year from the Tulare Basin and increase wastewater reuse by 695 taf per year (with no increase in seawater desalination). Ending Delta exports increases average water scarcity for Southern California customers by 260 taf per year, incurring an average scarcity cost to customers of $242 million per year. Central Coast cities supplied by the State Water Project are also shorted; they increase their wastewater reuse by 5 taf per year and seawater desalination by 48 taf per year. Urban water users in the Bay Area are able to adapt to the end of Delta exports with increased intertie capacity, including completion of EBMUD's Freeport project, an intertie between EBMUD and CCWD, and the Hayward Intertie between EBMUD and the Hetch Hetchy Aqueduct. In addition, more wastewater reuse (about 55 taf per year) is employed, as well as 187 taf per year of seawater desalination. There is also a 29 taf per year average reduction in water deliveries, incurring an average water scarcity cost for Bay Area users of $34 million per year.

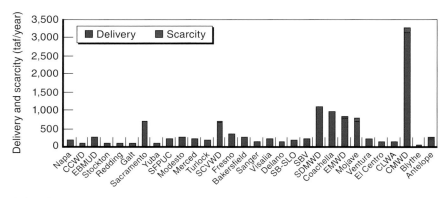

SOURCE: CALVIN model.

NOTES: CCWD, Contra Costa Water District; CLWA, Castaic Lake Water Agency; CMWD, Central Metropolitan Water District (including Los Angeles Department of Water and Power); EBMUD, East Bay Municipal Utilities District; EMWD, Eastern Metropolitan Water District; SCVWD, Santa Clara Valley Water District; SB-SLO, Santa Barbara–San Luis Obispo; SBV, San Bernardino Valley; SCVWD, Santa Clara Valley Water District; SDMWD, San Diego–Metropolitan Water District; SFPUC, San Francisco Public Utilities Commission.

Figure C.5—Annual Average Urban Water Deliveries and Scarcity by Urban Water Demand Area for 2050 Conditions with No Direct Delta Exports

The end of direct Delta exports reduces some pressure on environmental flows in the Sacramento Valley and Trinity River, commensurate with a reduced need to supply exports and Delta outflows. However, wetland water deliveries south of the Delta, represented as constraints in CALVIN, come with much higher costs to agricultural and urban water users. With exports, wetland water deliveries south of the Delta raise overall economic costs to agricultural and urban water users by an average of $20–$40 per acre-foot of environmental requirement. Ending direct Delta exports raises these average marginal opportunity costs to $90–$510 per acre-foot of environmental requirement. In other words, environmental water activities south of the Delta become considerably more expensive.

Regardless of the alternative (base case or no exports), the operating costs are significantly larger than the scarcity costs (Table C.2). Overall, operating costs amount to more than $3 billion per year (from pumping, water treatment, reuse treatment costs, desalination, etc.). In the no-

Table C.2

Average Annual Operating Costs With and Without
Delta Exports (2050 Water Demands) ($ millions)

Region	Base Case	No Exports	Cost Increase
Statewide	3,154	3,311	157
Sacramento Valley	195	206	12
San Joaquin Valley and Tulare Basin	998	974	−24
Southern California	1,961	2,131	169

exports alternative, statewide operating costs increased approximately $157 million per year. The reduction in pumping costs associated with the SWP and CVP is offset by increased desalination and wastewater reuse. Regionally, only the San Joaquin Valley and Tulare Basin areas saw decreased operating costs, mainly because of the reduction in pumping costs associated with the California Aqueduct and the Delta Mendota Canal. Operating costs increased modestly for the Sacramento Valley and more significantly for Southern California (mainly because of increased wastewater reuse and desalination).

The model also considers consumptive environmental requirements, such as wildlife refuge flows and required Delta outflows. The costs associated with the environmental flow requirements are affected by the state of the Delta export pumps. In the no-exports alternative, the marginal cost of the environmental flows increases south of the Delta and decreases north of the Delta (Table C.3). As in previous studies (Jenkins et al., 2001; Tanaka et al., 2006), the marginal cost of consumptive environmental requirements was higher than the nonconsumptive requirements. Consumptive environmental requirements cannot be used downstream for economic benefit; upstream environmental flows are typically nonconsumptive.

The marginal value of reservoirs and conveyance facilities indicates the per-acre-foot economic value that additional capacity would have for the statewide system. In general, there is greater value to increasing capacity for key conveyance facilities rather than reservoirs under the no-exports alternative (Table C.4). In most locations, the marginal value of additional reservoir capacity decreases without exports because there is less need to

Table C.3

Average Marginal Cost of Environmental Flow Requirements
($ per acre-foot)

Environmental flow Requirement	Region	Delta Exports	No Exports
Instream flow requirements			
Sacramento River	Sacramento Valley	1.2	1.5
San Joaquin River	San Joaquin Valley	8.8	90.0
Trinity River[a]	Sacramento Valley	34.8	31.7
Refuges			
Eastern Sacramento Valley refuges	Sacramento Valley	2.4	0.3
Western Sacramento Valley refuges	Sacramento Valley	2.7	0.4
Pixley National Wildlife Refuge	Tulare Basin	34.2	114.0
Kern National Wildlife Refuge	Tulare Basin	38.3	511.1
San Joaquin Wildlife Refuge	San Joaquin Valley	24.0	406.3
Other			
Mendota Pool	Tulare Basin	21.4	88.7
Required Delta outflow	Sacramento Valley	2.6	0.3

[a]Trinity River minimum instream flows are consumptive in CALVIN.

store water north of the Delta and there is less water to store south of the Delta. Reservoirs that would benefit from expansion tend to be in the Tulare Basin, where water can be exported to urban areas of Southern California. The maximum benefit of reservoir expansion would come from Lake Kaweah, but it would only be about $92 per acre-foot per year. North of the Delta, the value to increasing reservoir storage capacity is generally less than a $100 per acre-foot per year. In general, the changes in marginal values of expanding reservoirs increased only a small amount from the base case to the no-exports alternative (Lake Skinner was an exception, with a large decrease in value resulting from limited supplies to store).

Key conveyance facilities, on the other hand, would benefit from expansion. Facilities such as the Hayward Intertie, the Hetch Hetchy Aqueduct, the Colorado River Aqueduct, and the Mokelumne Aqueduct could provide additional benefits if expanded. These facilities would give urban areas in the Bay Area and Southern California access to more water, which becomes increasingly scarce without Delta exports. Facilities that provide water to the Bay Area (such as the Hetch Hetchy Aqueduct and the

Table C.4

Average Monthly Marginal Values of Expanded Capacity at Key Conveyance Facilities and Reservoirs ($ per acre-foot)

	Base Case	No Exports	Difference
Conveyance facilities			
Mokelumne River Aqueduct	0	112	112
Hetch Hetchy Aqueduct	193	608	415
New Don Pedro–Hetch Hetchy Aqueduct Intertie	170	583	409
Hayward Intertie	109	518	409
Cross Valley Canal	0	151	151
Friant-Kern Canal	0	2	2
Colorado River Aqueduct	169	488	319
Reservoirs			
Clair Engle Lake	0.2	0.2	0.0
Shasta Lake	0.5	0.4	−0.1
Lake Oroville	0.8	0.6	−0.2
Folsom Lake	0.7	0.6	−0.1
New Melones Reservoir	0.5	0.5	0.0
New Don Pedro Reservoir	0.5	0.4	−0.1
San Luis Reservoir	0.0	0.0	0.0
Millerton Lake	0.3	1.6	1.3
Lake Isabella	0.2	0.9	0.7
Lake Kaweah	2.9	9.3	6.4
Lake Success	2.6	8.3	5.7
Lake Skinner	29.4	1.5	−27.9

Hayward Intertie) are especially valuable because of larger scarcities in the urban areas without Delta exports.

Increased Minimum Outflow Scenarios

Current monthly net Delta outflow (MNDO) is approximately 5,593 taf per year, with the highest requirements in the spring and early summer months (Table C.5). For each modeling run in the increased minimum outflow scenarios, the monthly MNDOs were increased. For example, the "500" alternative in Figure C.6 means that the minimum Delta outflow

Table C.5

Current Required Monthly Net Delta Outflow
(taf per month)

Month	Minimum	Average	Maximum
October	246	260	354
November	208	259	268
December	215	267	277
January	277	345	594
February	374	775	1,581
March	369	871	1,713
April	349	753	1,432
May	238	689	1,713
June	304	548	1,431
July	246	396	537
August	184	251	336
September	179	179	179

required was 500 taf per month, i.e., all the monthly flows that were less than 500 are increased to 500.

As the minimum required outflows from the Delta increase, the surplus Delta outflows (those flows above the required volumes) decrease. Winter surplus flows remain high as a result of flood flows (high flow events) on the Sacramento River.

In these modeling scenarios, the minimum net Delta outflow was steadily increased, with smaller and less frequent periods of Delta inflows assumed to be available for export pumping. Monthly effects on net Delta outflows from October to September appear in Figure C.6. When minimum net Delta outflows are increased, less surplus Delta outflow (outflow in wet periods that exceeds outflow requirements and available storage capacity in upstream reservoirs) remains during winter months, and greater flows must be dedicated to outflows during summer and fall months. Higher levels of required outflow also imply reductions in water delivery upstream and in Delta exports.

A few assumptions regarding pumping apply in these scenarios. Banks Pumping Plant was assumed to have 8,500 cfs of hydraulic capacity. Currently, Banks is constrained to approximately 6,600 cfs because of

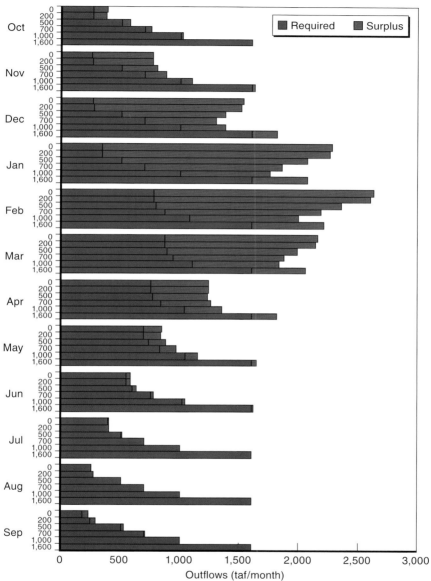

SOURCE: CALVIN model.
NOTES: For each month, the graph shows the increase in required and surplus
outflows, starting with current regulatory requirements ("0") and going up to 1.6 maf per
month.

**Figure C.6—Average Monthly Required and Surplus Net Delta Outflows with
Increasing Minimum Net Delta Outflows**

237

regulatory requirements, but as part of the South Delta Improvements Project, a program endorsed by CALFED, the pumping plant capacity would be increased to the hydraulic capacity. Tracy and the CCWD Pumping Plant capacities are unchanged from current conditions.

A major difference between the no-exports alternative and these increased-outflow cases is that water users employing eastern tributaries to the San Joaquin River can sell water to other water users south of the Delta by releasing water to the Delta, contributing to net Delta outflows and the availability of water in the Delta for pumping south (Figure C.7). It is economically optimal for both direct and indirect exporters of water from the Delta to share any required increases in Delta outflows. This is evident in comparing the amounts of water scarcity in Figure C.3 (no exports) to those in Figures C.8 and C.9 (increased net Delta outflows) for CVPM regions 11 and 12.

Some economically reasonable water transfers occur:

- Sacramento and in-Delta agricultural water users sell water south of the Delta (where the economic value of water value tends to be higher).
- Eastern San Joaquin Valley farmers (CVPM regions 11, 12, and 13) sell water to increase flows into the San Joaquin River and Delta (from which much is then exported).
- Western San Joaquin Valley and Tulare Basin farmers as well as Southern California urban areas purchase water from Eastern San Joaquin, Sacramento Valley, and in-Delta farmers.

As minimum net Delta outflows increase, scarcity for agricultural regions grows before urban regions experience any changes (Figure C.10). Agricultural users in both the Sacramento and San Joaquin Valleys have roughly the same rate of scarcity increase. Agricultural users in Southern California already transfer as much water as possible (given conveyance capacity constraints on the Colorado River Aqueduct) to the urban users. Thus, Southern California agricultural users are unaffected by changes in the required Delta outflows, except for higher prices paid for water they sell to Southern California cities. However, urban users in Southern California see increased scarcities before any other urban users.

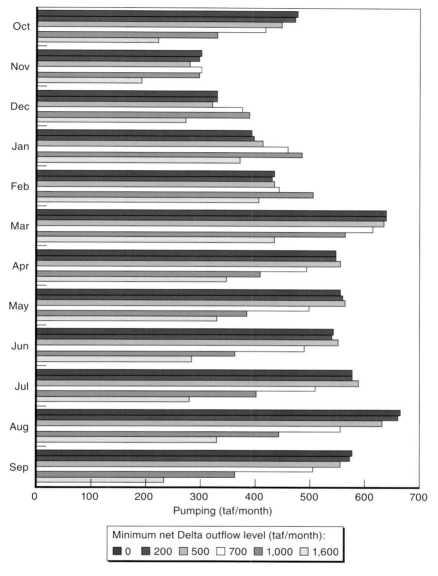

SOURCE: CALVIN model.

Figure C.7—Average Monthly Export Pumping for CVP and SWP with
Increasing Levels of Minimum Net Delta Outflows

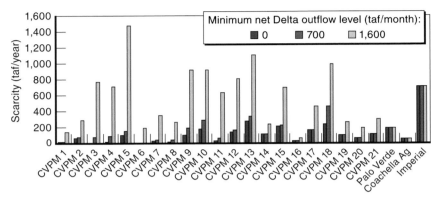

SOURCE: CALVIN model.

Figure C.8—Average Annual Agricultural Water Scarcity by Agricultural Area with Increasing Monthly Net Delta Outflow Requirements

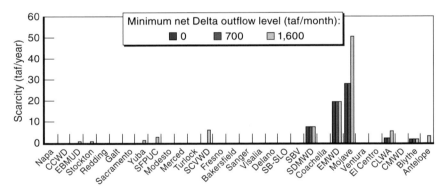

SOURCE: CALVIN model.
NOTES: For demand area abbreviations, see the notes to Figure C.5.

Figure C.9—Average Annual Urban Water Scarcity by Urban Area with Increasing Monthly Net Delta Outflow Requirements

Overall, the economic effects of increasing minimum net Delta outflows are less than the effects of eliminating Delta exports. Increases in water scarcity from higher Delta outflow requirements are shared among all users of waters flowing into the Delta, both direct and indirect Delta exporters. This additional flexibility greatly reduces the economic effects of increased Delta outflows and evens out the market values of water and

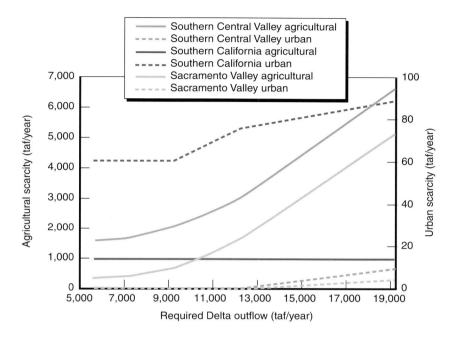

Figure C.10—Average Annual Regional Agricultural and Urban Water Scarcity with Increasing Net Delta Outflow Requirements

its opportunity costs for environmental, agricultural, and urban uses throughout California.

In these increased-outflow scenarios, operating costs remained relatively constant—around $3.1 billion per year—regardless of the required minimum net Delta outflow (Table C.6). Initially, the operating costs decrease as outflows are increased, but then costs increase beyond the initial levels at the highest outflow requirement level. Overall, as with the no-exports alternative, the base operating costs are far higher than the scarcity costs and incremental operating costs.

Unlike the no-exports alternative, in the increased-outflow scenarios, the marginal opportunity cost of environmental flow requirements grows throughout the state (Table C.7). The consumptive use requirements had the highest costs, ranging from $140 per acre-foot to $415 per acre-foot. Trinity River minimum instream flow requirements had the highest marginal cost. These flows are in the northernmost portion of the CALVIN system and are unavailable for use and reuse in the Sacramento

Table C.6

Regional Operating Costs with Increasing Net Delta Outflow Requirements
($ millions)

Region	Minimum Outflow (taf/month)					
	0	200	500	700	1,000	1,600
Statewide	3,153	3,153	3,152	3,143	3,117	3,170
Sacramento Valley	195	195	195	195	196	213
San Joaquin Valley and Tulare Basin	997	997	996	986	967	1,008
Southern California	1,961	1,961	1,961	1,961	1,955	1,949

Valley or south of the Delta. The minimum environmental flow requirements in the Sacramento and San Joaquin Rivers generate steadily increasing costs but are still an order of magnitude smaller than the other marginal costs. As with the no-exports alternative, in these scenarios the refuges south of the Delta have consistently higher marginal costs than refuges north of the Delta. The costs of the northern refuges increase significantly when urban water scarcity north of the Delta increases.

As with the no-exports alternative, in the increased-outflow scenarios, the greatest benefits would come from expanding conveyance facilities that provide water to urban areas (Table C.8). Initially, as the required net Delta outflows increase, the marginal value of increasing the capacity of the Bay Area facilities rises, but eventually it decreases when there is insufficient water to fill the existing capacity. The Colorado River Aqueduct would benefit from expansion. As Delta outflows are increased, urban Southern California would benefit from transferring more water from Colorado River agricultural users. The benefits of expanding the Cross Valley Canal and Friant-Kern Canal are relatively minor compared to those of the aqueducts and interties.

Summary

The Sacramento–San Joaquin Delta is the hub of the state's water resource system, with most of California relying on it, either directly or indirectly, for water. The State Water Project and Central Valley Project

Table C.7

Average Monthly Marginal Cost of Selected Environmental Flow Requirements with Increased Outflow Requirements ($ per acre-foot)

	Region	Minimum Outflow (taf/month)					
		0	200	500	700	1,000	1,600
Instream flow requirements							
Sacramento River	Sacramento Valley	1.7	1.7	2.1	2.4	4.6	33.7
San Joaquin River	San Joaquin Valley	8.1	8.2	9.3	10.9	14.5	14.5
Trinity River	Sacramento Valley	34.6	34.8	37.3	42.4	75.3	412.9
Refuges							
Eastern Sacramento Valley	Sacramento Valley	2.3	2.4	2.5	8.1	33.3	173.4
Western Sacramento Valley	Sacramento Valley	2.6	2.7	4.6	8.7	36.0	131.1
Pixley National Wildlife	Tulare Basin	33.2	33.3	35.6	39.5	67.6	113.0
Kern National Wildlife	Tulare Basin	37.4	37.5	39.8	44.6	77.3	151.7
San Joaquin Wildlife	San Joaquin Valley	23.2	23.3	25.8	30.9	62.4	361.7
Other							
Mendota Pool	Tulare Basin	20.6	20.7	22.8	26.8	50.7	277.0
Required Delta outflow	Sacramento Valley	2.5	2.8	4.9	9.4	39.0	339.3

directly export water from the Delta for Southern California and Bay Area cities and San Joaquin and Tulare Basin irrigation, respectively. Local urban water districts and in-Delta agriculture also rely on withdrawals from the Delta to meet their water needs. Upstream of the Delta, irrigation and urban users withdraw water from the major rivers and tributaries that would have otherwise flowed into the Delta. Changes in operations or Delta outflow requirements can significantly affect the availability of water from the Delta. CALVIN model results cannot provide us with an exact map of how to manage the Delta, nor can they predict the exact costs associated with changes in operations or requirements. However, they can provide insight into operations and costs associated with major changes.

Table C.8

Average Marginal Value of Expanding Key Conveyance Facilities Under Increased-Outflow and No-Export Conditions ($ per acre-foot)

Region	Monthly Net Delta Outflow (taf/month)						No Exports
	0	200	500	700	1,000	1,600	
Mokelumne River Aqueduct	0	0	0	0	2	18	112
Hetch Hetchy Aqueduct	255	210	184	237	201	183	608
Hayward Intertie	109	109	107	106	106	102	409
Cross Valley Canal	0	0	0	0	0	1	151
Friant-Kern Canal	0	0	0	0	1	1	2
Colorado River Aqueduct	137	142	172	139	169	208	319

Appendix D

Delta Agricultural Production Model

Estimation of economic effects of water policies and management to agriculture has become commonplace in recent decades. The most common approach is to develop mathematical models of farmer behavior that assume that farmers behave like businesspeople. That is, they aim to maximize their business profits, given agricultural commodity prices, within the boundaries of agricultural production functions, availability, cost, effectiveness of irrigation technologies, and limitations of available land, water, and capital resources (Howitt, 1995; Howitt, Ward, and Msangi, 2001). The DAP model presented below provides such a model for the Delta, disaggregated by agricultural islands and including the effects of salinity in the water supply on crop yields. The model is similar to the CVPM, CALAG, and Statewide Water and Agricultural Production (SWAP) models, which are commonly used for modeling agricultural land and water use and economic performance throughout California (U.S. Bureau of Reclamation, 1997; Howitt, Ward, and Msangi, 2001; Howitt, Tauber, and Pienaar, 2003).

Farmers often adapt to changes in crop price, land availability, and water availability and quantity. The DAP model provides a reasonable indication of how farmers are likely to adapt to changes in water availability, water quality, and, particularly local salinity. Although the model assumes that farmers make decisions as businesspeople do in response to such changes, it does not include some adaptation options available to farmers, such as modifying their irrigation practices to avoid saltier seasons or parts of tidal cycles, or selecting less salty sides of Delta islands for water withdrawals. As such, the results presented in Chapter 6—which show a 10 percent overall decline in crop revenues with a tenfold increase in salinity—are conservative estimates of the losses of revenues associated with increases in salinity. These losses could be lower if farmers were able to make some of these adaptations. Delta farm production is small relative to the total market of produced commodities, so changes in Delta production seem unlikely to greatly affect market commodity prices.

Model Formulation

This model was built as an extension of the existing SWAP model of statewide agricultural production, which uses positive mathematical programming to calibrate the production function (Howitt, 1995). In the SWAP model, the Sacramento–San Joaquin Delta is represented by parts of two agricultural regions.[1] The model presented herein contains a more disaggregated representation of the Delta and it also incorporates the effect of water salinity on agricultural production. The model includes the following steps (adapted from Howitt and Msangi, 2002):

Step 1: Calibration using linear programming

$$\text{Max} \sum_{g} \sum_{j} p_{gj} y_{gj} x_{gj,\text{land}} - \sum_{g} \sum_{j} \sum_{i=\text{water,labor}} \omega_{gji} x_{gji} a_{gji}$$

$$
\begin{array}{llll}
\text{s.t.} & Ax \leq b & \forall g,j,i & \text{(resource constraint)} \\
& Ix \leq \tilde{x} + \varepsilon & \forall g,j \quad i = \text{land, water} & \text{(upper bound calibration on land)} \\
& Ix \leq \tilde{x} - \varepsilon & \forall g,j \quad i = \text{land, water} & \text{(lower bound calibration on land)} \\
& x \geq 0 & \forall_{g,j,i} & \text{(nonnegativity constraint)}
\end{array}
$$

where p_{gj}, y_{gj}, and x_{gj} are price, yield, and land, respectively, of crop j in region g; ω_{gji} and a_{gji} are, respectively, cost and Leontieff coefficient for input i (labor, water, and land) for crop j in region g. The matrix A in the resource constraint contains the Leontieff coefficients a_{gji}.

Step 2: Cost function parameters

Consider a quadratic cost function

$$TC_{gj} \alpha_{gj} x_{gj,\text{land}} - \frac{1}{2} \gamma_{gj} x_{gj,\text{land}}^2$$

where α_{gj} and γ_{gj} are the intercept and the slope of a marginal cost function for input i crop j in region g. In the empirical model, only the land cost function has nonzero γ_{gj} values. The dual value on land, λ, is

---

[1] In SWAP and CALVIN, Delta agriculture is included in regions 6 and 9, depicted in Figure C.2 (Appendix C).

obtained from step 1 for each crop and region. The parameters for the land cost function are obtained as follows, where η_j is j's output elasticity of supply:

$$\gamma_{gj} = \frac{P_{gj}y_{gj}^2}{\eta_j x_{gj}} \quad \text{and} \quad \alpha_{gj} = \omega_{gj,\text{land}} + \lambda_{gj} - \gamma_{gj} x_{gj,\text{land}}$$

Step 3: Nonlinear profit maximization

The final step is the following nonlinear profit maximization program, which considers a constant elasticity of substitution (CES) production specification for every crop and every region:

$$\text{Max} \sum_g \sum_j P_{gj} Y_{gj} Yr_{gj} - \sum \sum \sum (\alpha_{gji} X_{gji} - \gamma_{gji} X_{gji}^2)$$

s.t. $AX \leq b$

$$Y_{gj} = \tau_{gj} \left(\sum_i \beta_{gji} X_{gji}^{\frac{\sigma}{\sigma-1}} \right)^{\frac{\sigma-1}{\sigma}}$$

where Y_{gj} is the output level of crop j in region g. τ_{gj} is a scale factor, and β_{gji} is a share parameter for input i. The elasticity of substitution, σ is assumed constant for all crops and regions.

The effect of salinity on agricultural production is represented by the relative yield Yr_{gj}, proposed by van Genuchten and Hoffman (1984).

$$Yr_{gj} = \frac{1}{1 + \left(C_g / C_{50j}\right)^2}$$

where C_g is the root zone salinity in region g and C_{50j} is the root zone salinity at which the yield of crop j reduces by 50 percent (Table D.1). A graphical representation of the van Genuchten equation is presented in Figure D.1.

Model Regions

The model regions were defined considering two criteria: the agricultural land allocation data available and the spatial distribution of

Table D.1

Root Zone Salinity Levels That Reduce Yields by 50 Percent

Crop	C_{50} (mS/cm)
Alfalfa	6.85
Field corn	6.85
Grain	13.04
Orchard	4.13
Pasture	8.85
Rice	18.00
Sugar beet	13.04
Tomato	6.85
Truck crop	6.50
Wine grape	8.85

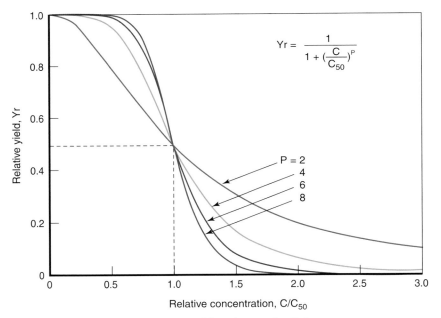

$$Yr = \frac{1}{1 + (\frac{C}{C_{50}})^P}$$

P = 2
4
6
8

Relative yield, Yr

Relative concentration, C/C_{50}

Figure D.1—Yield Reduction by Salinity

salinity in the Delta. Delta land use was obtained from the California Department of Water Resources 1990s surveys. The data are disaggregated into the 72 spatial units presented in Figure D.2, panel a. The blank area depicts land that cannot be included in the model because the land use data are not sufficiently detailed.[2] Land uses are categorized as urban, agricultural, and environmental. Agricultural use includes disaggregated values for the following crop categories: pasture, alfalfa, field corn, sugar beets, grain, rice, truck crops, tomato, orchards, and vineyards. Environmental use includes native riparian, water surface, and native vegetation.

Units with less than 3 percent agricultural land were excluded from the model. This includes Bethel Tract, Sargent-Barnhart Tract, Browns Island, Chipps Island, Clifton Court Forebay, Fay Island, Kimball Island, Little Franks Tract, Little Mandeville Island, Little Tinsley Island, Mildred Island, Neville Island, Rhode Island, Van Sickle Island, Winter Island, Sycamore Island, and the undesignated islands. All these units are predominantly environmental, with the exception of Sargent-Barnhart Tract, which is urban.

To reduce the computational effort in the model calibration, some islands were grouped into aggregated agricultural regions. Specifically, this was done for the northern part of the Delta, where salinity concentrations are low and relatively uniform. Therefore, the model includes 35 regions, 33 of which correspond to original units in the DWR data. The two northern regions are aggregations of single units, under the jurisdiction of the North Delta Water Agency (Northwest) and the Central Delta Water Agency (Northeast). The model regions are shown in Figure D.2, panel b. Hatched areas are not included in the model—this includes the Delta water channels or other areas of open water (such as Franks Tract), areas excluded because of land use data problems (the blank areas in Figure D.2, panel a), and areas with less than 3 percent agricultural land, as noted above. Of the 460,000 agricultural acres in DWR surveys from the 1990s, 332,400 acres (72%) are represented in the model. The remaining agricultural areas are

---

[2]Within this area, salinity levels vary, but the land use dataset treats the entire area as a single unit.

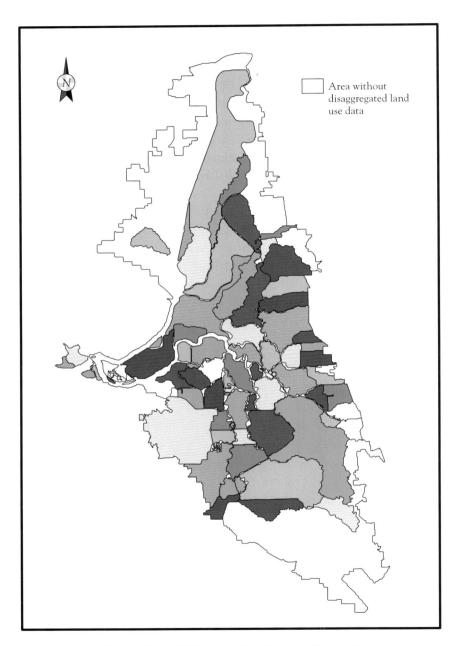

Area without
disaggregated land
use data

Figure D.2—DAP Modeling Regions (Panel a)

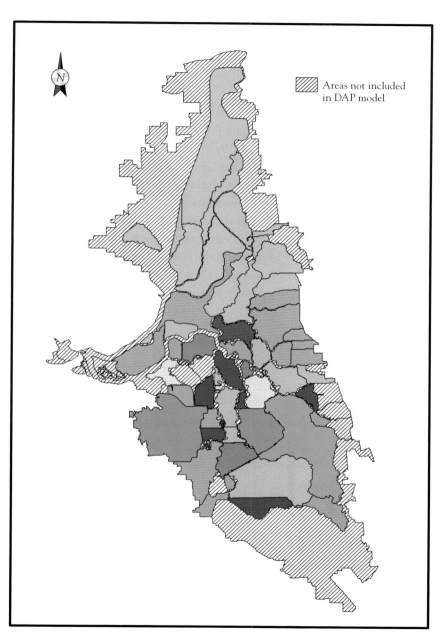

Areas not included
in DAP model

Figure D.2—DAP Modeling Regions (Panel b)

predominantly in upland parts of the Delta, which do not receive water supplies directly from the Delta's interior or western edge and which are less likely to be affected by changes in Delta salinity.

Spatial Distribution of Salinity

Electrical conductivity (EC) data for 19 Delta locations were obtained from the California Department of Water Resources web site. The common period of record available for all monitoring stations of interest was from August 1999 to May 2006. Figure D.3 shows the average salinity over the irrigation season (July to September) at each monitoring point.

It can be observed that most of the stations have an EC less than 1 mS/cm, which in practice means no effect on agricultural production. The three stations with the highest salinities are, as expected, at the west extreme of the Delta. The station with the highest salinity (Pittsburg) has an EC around 12 percent of seawater salinity (~45 mS/cm). Using the values presented in Figure D.3, each model region was assigned a value of salinity.

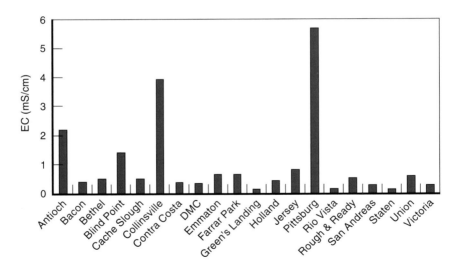

Figure D.3—Average Salinity Levels over the Irrigation Season at Delta Monitoring Points

Modeling Scenario

The historical salinity distribution was considered as the model base case. Two additional salinity scenarios were explored. The spatial distribution of salinity for these scenarios was obtained by scaling the base salinity distribution by factors of 10 and 20, respectively. Chapter 6 reports the changes in crop revenues and profitability with a tenfold salinity increase. The corresponding results for a twentyfold salinity increase are shown in Figure D.4. In this scenario, overall crop revenues and profits in the Delta are reduced by about one-third (to $254 million per year and $135 million per year, respectively). Table D.2 shows the acreages devoted to different crops under each scenario. Overall crop acreage declines by about 2 percent and 10 percent, respectively, in the ten- and twentyfold scenarios, and there are some shifts from higher-value fruits and vegetables toward field crops and pasture as salinity rises.

Conclusions

This initial version of the Delta Agricultural Production model provides a tractable way to estimate the effects of different Delta management scenarios on the agricultural and agricultural economic performance of the Delta. Its application to a base case representing current salinity conditions and to a set of higher Delta salinities illustrates the potential value of results from such a model. The results also indicate that substantially higher Delta salinities do not necessarily bring an end to agriculture in the Delta, although they are accompanied by substantial losses of agricultural revenues and profit.

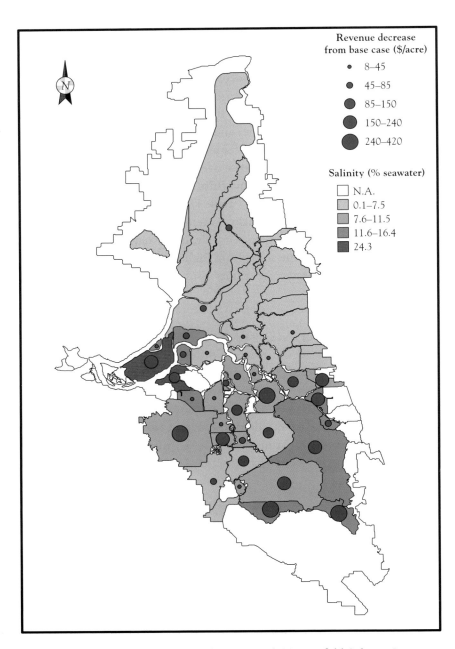

Figure D.4—Map of Revenue Reductions with Twentyfold Salinity Increase

Table D.2

Crop Acreages in the Delta Under Base Case Conditions and with Increased Salinity Levels

	Field Crops and Pasture					Fruits and Vegetables				Total
	Alfalfa	Grains	Sugar Beets	Other Field Crops	Pasture	Tomatoes	Truck Crops	Orchards	Wine Grapes	
Base case (summer)										
North	11,200	22,500	4,400	60,800	4,200	13,500	3,300	8,200	12,600	140,500
Central	6,000	13,500	700	38,700	500	2,100	16,000	300	1,600	79,400
West	3,900	5,300	200	13,800	6,700	4,300	2,400	5,600	200	42,400
South	19,200	7,700	1,600	18,900	400	8,000	11,500	1,200	1,600	70,000
Total	40,200	49,000	6,900	132,100	11,900	28,000	33,100	15,300	16,000	332,400
Tenfold salinity increase										
North	11,000	22,600	4,400	59,100	3,900	13,200	3,200	7,800	12,400	137,500
Central	5,900	14,100	700	38,600	600	2,100	15,400	300	1,600	79,200
West	3,800	5,500	200	13,800	6,600	4,200	2,300	4,900	200	41,500
South	18,500	8,600	1,600	18,800	400	7,500	10,500	900	1,500	68,400
Total	39,200	50,800	6,900	130,300	11,500	26,900	31,300	13,800	15,700	326,500
Twentyfold salinity increase										
North	10,400	22,100	4,300	55,300	3,100	12,400	3,000	6,700	12,000	129,200
Central	5,700	14,500	700	37,400	500	2,000	13,900	200	1,600	76,300
West	3,500	5,700	200	12,900	5,700	3,600	2,000	3,600	200	37,400
South	15,200	8,100	1,500	15,600	400	6,100	8,400	600	1,300	57,200
Total	34,700	50,300	6,700	121,100	9,600	24,100	27,200	11,200	15,000	300,100

Appendix E
Cost Elements for Delta Alternatives

The preliminary cost estimates presented in this report allow rough conceptual comparison of Delta alternatives. Although they illustrate the types of comparisons possible, a full evaluation of Delta alternatives will require more complete, accurate, and detailed estimates. This appendix details the basis for our cost estimates.

Costs considered here include investment costs for capital infrastructure, land purchase costs for rights-of-way, water scarcity costs (economic losses to water users from reductions in water deliveries), and operating costs for pumping and water treatment. We consider costs to state, federal, and local water agencies, as well as to individual water users. Capital and operating costs for additional fish screening or bank filtration actions to prevent fish and fish larvae entrainment are not considered. To adjust cost estimates from earlier studies to 2006 dollars, we use the *Engineering News Record* construction cost index. Larger macroeconomic effects (both costs and benefits) were beyond the scope of this work. We also have not included cost contingencies. As has been seen with several major recent infrastructure projects in California, we are likely to see surprises between preliminary and final cost estimates and the final cost of any completed project.[1]

Investment and Other Cost Estimates

Levees as Usual. The investment costs of this scenario assume an increased level of effort relative to that in the recent past. A CALFED study from the late 1990s estimated that it would cost roughly $1 billion to bring Delta levees up to PL 84-99 standards (the federal standard for

---

[1]See Flyvbjerg, Bruzelius, and Rothengatter (2003) for a more general discussion of cost overrun issues.

100-year agricultural levees). This exercise assumed that not all of the 1,100 miles of levees required the same level of effort. More recent informal estimates by water managers have been in the range of $2 billion, assuming roughly $2 million per linear mile for levee upgrades to these standards. In some locations, we are aware of detailed cost estimates as high as $16 million per mile and as low as $1.4 million per mile.

Fortress Delta. Recent informal estimates by water managers of the cost of significantly fortifying levees—including significant structural work—have been on the order of $5 million per linear mile. Dutch levels of levee protection are considerably higher and would probably involve changes in many islands and channels, straining current construction and levee material capacity. So we increase this cost estimate to $10 million per mile. For such a large and fundamental reengineering and upgrading of the Delta levee system, this estimate is necessarily rather speculative. Our estimate of capital costs for this alternative (upward of $3 billion to $5 billion) assumes that 300 to 500 miles of levee would be fortified to Dutch levels, with other levees incurring additional structural expenses.

Seaward Saltwater Barrier. The most detailed cost estimates of a saltwater barrier date back to investigations done in the late 1920s and early 1930s (Young, 1929; Matthew, 1931b), and they ranged from $40 million (for a barrier at Chipps Island) to $75 million (for a barrier at Point San Pablo). Carrying forward these estimates to today's values with standard engineering cost deflators, such a barrier would now cost on the order of $1.7 billion to $3.2 billion to build. This range is likely to be on the low side, given the additional costs of modern regulatory review requirements. The Maeslant movable barrier near Rotterdam in The Netherlands was completed in 1997 at a cost of over $800 million. The larger fixed Eastern Scheldt storm surge barrier in The Netherlands was completed in 1987 at a cost of over $3 billion (see www.deltawerken.com).

Peripheral Canal Plus. The most recent cost estimate for a peripheral canal was produced for CALFED (CALFED, 1999). For a 10,000 cfs incised earthen canal complete with fish screens, drainage, siphon, and control structures, it assumed a total capital cost of $1.9 billion in 1998 dollars ($2.5 billion in 2006 dollars).[2] In current discussions of a peripheral canal, the question of land costs often comes up. The 1999 study that assumed 6,000 acres of land purchase would be required for right-of-way, at an average price of $3,500 per acre ($4,550 per acre in 2006 dollars). Costs might be $120 million higher if this underestimates the share of urban or urbanizing lands that would need to be purchased.[3] According to DAP results, allowing extreme salt intrusion into the western, central, and eastern Delta under this alternative would decrease agricultural land use by 109,000 acres, revenues by $119 million per year, and profits by $70 million per year. Our cost estimate for the Peripheral Canal Plus alternative—in the range of $2 billion to $3 billion—does not include costs for Delta ecosystem support, selected urban levee improvements, and possibly also some other levees or channel modifications to prevent deterioration of water quality within the Delta that would accompany this program.

South Delta Restoration Aqueduct. Because this proposal is new, there are no previous studies from which we can draw for investment cost estimates. Our estimate of several billion dollars reflects the fact that many costs are likely to be comparable to those of the Peripheral Canal Plus. According to DAP results, allowing extreme salt intrusion into the western and central Delta under this alternative would decrease agricultural land use by 68,000 acres (about 57% of current farmland in the western and central Delta), agricultural revenues by $70 million per year, and agricultural profits by $41 million per year. This is a likely upper bound to agricultural losses within the Delta for this alternative.

Armored-Island Aqueduct. Variants of this alternative have been examined and costs estimated by CALFED (1997) and Orlob (1982). Capital costs for Orlob's through-Delta proposal were estimated in the

---

[2]Studies from the late 1970s and early 1980s estimated significantly lower costs for a raised canal that would have been over twice as large (22,000 cfs): $1.6 billion to $1.8 billion in 2006 dollars (Orlob, 1982).

[3]Agricultural lands in the eastern Delta currently sell for $2,000 to $3,000 per acre, but lands slated for development can sell for $10,000 per acre or more. If the canal's trajectory could not avoid some already developed land, some acreage would sell at much higher prices. The $120 million additional cost estimate assumes average land costs of roughly $20,000 per acre.

range of $330 million to $545 million ($0.8 billion to $1.1 billion in 2006 dollars). The CALFED analysis included two through-Delta alternatives, ranging in cost from $0.8 billion to $1.4 billion in 1997 dollars ($1.1 billion to $1.9 billion in 2006 dollars). Our cost estimate of $1 billion to $2 billion assumes that this range is still appropriate for this type of investment. The need to build a system for ship passage to the Port of Stockton, for instance with operable gates, might increase costs above this range. The estimate does not include additional investments for urban levees and environmental programs. DAP results show that if Delta islands west of such an aqueduct were eliminated from agricultural use, as a worst case, the loss from agriculture would be about 46,000 acres, $47 million in agricultural revenues, and $27 million in agricultural profits. Similar losses could occur for other alternatives that would make the western Delta more saline.

Opportunistic Delta. This alternative assumes that capital costs will largely be concentrated on additional storage in the vicinity of the Delta, to allow large amounts of water to be stored when flows are high and then released into aqueducts as conveyance capacity becomes available. Recent estimates by the CALFED surface storage investigations team (CALFED, 2006) put the capital costs of Los Vaqueros Expansion at $0.9 billion to $1.5 billion (for an expansion of storage capacity by 200–400 taf) and the costs of using two Delta islands (Webb and Bacon) as storage at $0.7 billion to $0.8 billion (for 217 taf of storage). Our range of near-Delta storage cost estimates ($0.7 billion to $2.2 billion) allows for storage investments over the range available (217–617 taf). Other investments, including more groundwater storage and recycling south and west of the Delta, would also be required. Recycling (capital and operating) and groundwater (operating) costs are included in the CALVIN results shown in Chapter 6 and Appendix C. Additional costs for capital facilities would likely be incurred. Water scarcity costs should increase by less than $170 million per year ($120 million in statewide costs, according to CALVIN, and less than $50 million in lost agricultural profits according to DAP results), and might be less than $50 million per year with additional near-Delta storage.

Eco-Delta. The several billion dollars in capital costs for this alternative would allow for investments to use Delta islands for various purposes, as described in Chapters 4 and 7 (see especially Figure 7.6).

Some water user capital investments would take place, although near-Delta storage might be less attractive because it would be filled less frequently. Existing storage and conveyance south of the Delta might be sufficient, given much lower typical values of pumping. Water supply infrastructure costs are likely to be less than those for the Abandoned Delta alternative. Water scarcity costs, according to CALVIN results, should increase by less than $500 million per year. Some additional costs for wastewater reuse would also occur, perhaps for some seawater desalination as well. Losses of half the agricultural profits in the core Delta area modeled by DAP would amount to $100 million per year. (A twentyfold increase in Delta salinity reduces profits by about a third.) Total annual costs should be less than $600 million per year.

Abandoned Delta. In this alternative, there are no capital investments within or near the Delta, but water users make investments in interties and alternative sources (groundwater banking, recycling, desalination). Desalination and recycling costs are included in the CALVIN operating cost estimates shown in Chapter 6 and Appendix C. Many of the interties are already being constructed for various reasons. We estimate the additional costs to be on the order of $500 million. Additional water scarcity and operating costs are estimated by CALVIN at about $1 billion per year, with an additional loss of up to $200 million per year in agricultural profits in the Delta as islands fail, for annual costs of $1.2 billion per year.

The cost of fortifying urban levees to levels of protection exceeding 200-year average recurrence could be added to several of the above alternatives (including Fortress Delta, if urbanization occurs behind levees that are not targeted as part of the basic investment program required to keep water supplies fresh). This action could cost $0.2 billion to $1.5 billion, depending on the length of urban levees and the level of protection sought. The lower estimate assumes $2 million per mile and 100 miles of urban levees and the higher estimate assumes $10 million per mile for 150 miles of levees designed to a Dutch standard. These estimates do not include flood control costs for urban and urbanizing areas outside the Delta, such as the Sacramento, Marysville-Yuba, and Modesto metropolitan areas.

Water Scarcity Costs

To assess water scarcity costs, we draw on the analysis of water user adjustment in Chapter 6. That chapter relied on CALVIN to look at the costs of reducing or eliminating exports and on DAP to look at costs to Delta agriculture of reductions in Delta water quality (increases in salinity).

As seen in Table 8.3, most alternatives continue to provide a range of Delta exports comparable to recent levels (six million acre-feet per year or more). The Reduced-Exports alternatives generally cut back on this range, entailing adjustment costs for water exporters. Estimates of these water scarcity costs—expressed as yearly costs—are calculated from the CALVIN runs presented in Chapter 6. The Abandoned Delta alternative uses the estimates for the no-export CALVIN scenario, and the Opportunistic Exports and Eco-Delta alternatives use estimates from the increased-outflow model runs that raised minimum net outflows (see Appendix C).

For Delta farmers, only the alternatives that continue to keep the entire Delta fresh would allow agriculture in the whole Delta, although some agricultural land would go out of production because of island flooding. The Fluctuating Delta alternatives (4–6) and the Opportunistic Delta (7) would all be broadly comparable to the scenario of increasing salinity analyzed in Chapter 6: Delta islands to the north and east would likely continue to have access to sufficiently fresh water to continue farming profitably, but some islands would go out of production in the west, center, and south. The costs of salinity-related reductions in agricultural production would differ with details of the alternative, but we can estimate them in a range of $38 million to $200 million per year, which seems likely to be an overestimate, since the most affected areas tend to have the lowest economic values for farm production. The areas of the Delta not represented in DAP tend to be upland and mainland areas that do not receive water from the western and central Delta. Under the Eco-Delta alternative (8), these costs might be somewhat higher, because farming activities would be tailored to ecosystem needs (e.g., corn rather than reduced acreage) and eco-friendly methods (restorative of the soils). The Abandoned Delta alternative (9) would see a phase-out of Delta farming, at a cost of up to $367 million per year in forgone farm revenues and up to $201 million per year in profits. Comparable costs might occur in a Levees-as-Usual alternative under the most catastrophic levee failure scenario.

Bibliography

Aguado, E., D. Cayan, L. Riddle, and M. Roos, "Climatic Fluctuations and the Timing of West Coast Streamflow," *Journal of Climate*, Vol. 5, 1992, pp. 1468–1483.

Alexander, B. S., G. H. Mendell, and G. Davidson, *Report of the Board of Commissioners on the Irrigation of the San Joaquin, Tulare, and Sacramento Valleys of the State of California,* Government Printing Office, Washington, D.C., 1874.

Anderson K., *Tending the Wild: Native American Knowledge and the Management of California's Natural Resources,* University of California Press, Berkeley, California, 2005.

Anonymous, "Regulatory Commitments—User Contributions. Statement of Principles," University of the Pacific negotiation group on an HCP-NCCP proposal for the Delta, Stockton, California, December 20, 2005.

Arax, Mark, and Rick Wartzman, *The King of California: J. G. Boswell and the Making of a Secret American Empire,* Public Affairs, New York, 2003.

Arnett, G. Ray, "Maintenance of Fish and Wildlife in the Sacramento–San Joaquin Estuary in Relation to Water Development," California Department of Fish and Game, Sacramento, California, April 1973.

Associated Press, "Old Delta Water Plan Resurfaces," December 14, 2004.

Association of California Water Agencies, *No Time to Waste: A Blueprint for California Water,* Sacramento, California, October 2005.

Atwater, B. F., *Geologic Maps of the Sacramento–San Joaquin Delta*, U.S. Geological Survey, MF-1401, Menlo Park, California, 1982.

Atwater, B. F., S. G. Conard, J. N. Dowden, C. W. Hedel, R. L. Donald, and W. Savage, "History, Landforms, and Vegetation of the Estuary's Tidal Marshes, in T. J. Conomos, A. E. Leviton, and M. Berson, eds., *San Francisco Bay: The Urbanized Estuary,* AAAS, Pacific Division, San Francisco, California, 1979, pp. 347–385.

Barbassa, Juliana, "Board Orders Enforcement of Delta Salt Limits; Agencies Told to Meet Rules or Face Fines and Shutdown of Pumps," *Modesto Bee*, February 20, 2006.

Baumol, W. J., and R. D. Willig, "Fixed Costs, Sunk Costs, Entry Barriers, and Sustainability of Monopoly," *Quarterly Journal of Economics,* Vol. 96, No. 3, August 1981, pp. 405–431.

Bay Institute, "From the Sierra to the Sea: The Ecological History of the San Francisco Bay–Delta Watershed," 1998, available at http://www.bay.org/sierra_to_the_sea.htm.

Bennett, W. A., "Critical Assessment of the Delta Smelt Population in the San Francisco Estuary, California," *San Francisco Estuary and Watershed Science,* Vol. 3, No. 2, 2005, pp. 1–71, available at http://repositories.cdlib.org/jmie/sfews/vol3/iss2/art1.

Bennett, W. A., and P. B. Moyle, "Where Have All the Fishes Gone: Interactive Factors Producing Fish Declines in the Sacramento–San Joaquin Estuary," in J. T. Hollibaugh, ed., *San Francisco Bay: The Ecosystem,* AAAS, Pacific Division, San Francisco, California, 1996.

Boxall, Bettina, "Delta Plan Is Dealt a Blow," *Los Angeles Times,* October 11, 2005.

Boxall, Bettina, "Repeat of Tragedy Feared in San Joaquin Drainage Plan," *Los Angeles Times,* July 8, 2006.

Braeutigam, R. R., "Optimal Pricing with Intermodal Competition," *American Economic Review,* Vol. 69, No. 1, March 1979, pp. 38–49.

Breitler, A., "Giant Gate for Delta? Katrina Disaster Renews Interest in European Barriers," *Stockton Record,* October 12, 2006.

Brown, L. R., "Will Tidal Wetland Restoration Enhance Populations of Native Fishes?" *San Francisco Estuary and Watershed Science,* Vol. 1, No. 1, 2003, available at http://repositories.cdlib.org/jmie/sfews/vol1/iss1/art2.

CALFED, *Workshop 5 Information Packet: Draft Alternatives,* CALFED Bay-Delta Program, Sacramento, California, February 1996.

CALFED, "Storage and Conveyance Component Inventories, Preliminary Working Draft," Sacramento, California, March 7, 1997.

CALFED, *Isolated Facility: Conceptual Analysis of Incised Canal Configuration,* Sacramento, California, September 1999.

CALFED, *Programmatic Record of Decision,* Sacramento, California, 2000a.

CALFED, *Final Programmatic Environmental Impact Statement/ Environmental Impact Report,* Sacramento, California, 2000b.

CALFED, *California's Water Future: A Framework for Action,* Sacramento, California, June 2000c.

CALFED, "Bay-Delta Authority Adopts Delta Improvements Package. Actions Will Protect Environment, Improve Water Supply Reliability," news release, Sacramento, California, August 13, 2004a.

CALFED, *Finance Plan,* Sacramento, California, December 2004b.

CALFED, *Surface Storage Investigations Progress Report,* Sacramento, California, May 2006.

California Urban Water Agencies, "Cost Allocation Issues for Long-Term Delta Investments," White Paper, Sacramento, California, March 27, 2006.

California Water and Environmental Modeling Forum, *A Strategic Analysis Framework for Managing Water in California,* Report 2005-1, Sacramento, California, September 2005.

Carlton, J. T., J. K. Thompson, L. E. Schemel, and F. H. Nichols, "Remarkable Invasion of San Francisco Bay (California USA) by the Asian Clam, *Potamocorbula amurensis.* I. Introduction and Dispersal," *Marine Ecology Progress Series,* Vol. 66, 1990, pp. 81–94.

Chessie System Railroads, Consolidated Rail Corporation, Family Lines Rail System, Norfolk and Western Railway Company, and Southern Railway System, "Comments of Five Railroads," Ex Parte 347 (Sub-No. D), Vol. I, May 20, 1981, pp. 111–112.

Cohen, A. N., and J. T. Carlton, "Accelerating Invasion Rate in a Highly Invaded Estuary," *Science,* Vol. 279, 1998, pp. 555–557.

Cohen, A. N., and A. Weinstein, "The Potential Distribution and Abundance of Zebra Mussels in California," a report for CALFED and the California Urban Water Agencies, San Francisco Estuary Institute, Richmond, California, 1998.

Contra Costa Times, "Editorial: Time to Repair Delta's Levee System," February 26, 2006.

Contra Costa Water District, *Urban Water Management Plan,* Concord, California, 2005.

Cooper, Audrey, "Historic Delta Water Deal Near," *San Joaquin Record,* October 29, 2003.

Crain, P. K., K. Whitener, and P. B. Moyle, "Use of a Restored Central California Floodplain by Larvae of Native and Alien Fishes," in F. Feyrer, L. R. Brown, R. L. Brown, and J. J. Orsi, eds., *Early Life History of Fishes in the San Francisco Estuary and Watershed,* American Fisheries Society Symposium 39, Bethesda, Maryland, 2004, pp. 125–140.

Craine, J., "Paradigms Undefined. Review of Ecological Paradigms Lost: Routes of Theory Change by B. Beisner and K. Cuddington," *Bioscience,* Vol. 56, 2006, pp. 447–449.

Damus, S., "Two-Part Tariffs and Optimum Taxation: The Case of Railway Rates," *American Economic Review,* Vol. 71, No. 1, March 1981, pp. 65–79.

de Alth, Shelley, and Kim Rueben, *Understanding Infrastructure Financing for California,* Occasional Paper, Public Policy Institute of California, San Francisco, California, June 2005.

Department of Boating and Waterways, *Sacramento–San Joaquin Delta Boating Needs Assessment,* Sacramento, California, 2002.

Department of Finance, *A Fiscal Review: CALFED Bay-Delta Program Summary of Expenditures as of September 30, 2004,* Office of State Audits and Evaluations, Sacramento, California, October 2005.

Department of Fish and Game, *State and Federally Listed Endangered and Threatened Animals of California,* Biogeographic Data Branch, California Natural Diversity Database, Sacramento, California, July 2006a.

Department of Fish and Game, "Yolo Basin Wildlife Area Draft Land Management Plan," Sacramento, California, October 2006b, available at http://www.yolobasin.org/management.cfm.

Department of Public Works, *Water Resources of California,* Public Works Bulletin No. 4, a Report to the Legislature of 1923, Sacramento, California, 1923.

Department of Public Works, *The State Water Plan*, Public Works Bulletin No. 25, a Report to the Legislature of 1931, Sacramento, California, 1930.

Department of Water Resources, *California Water Plan,* Bulletin 3, Sacramento, California, 1957.

Department of Water Resources, *Sacramento–San Joaquin Delta Atlas,* Sacramento, California, 1995.

Department of Water Resources, *California Water Plan Update,* Bulletin 160-98, Sacramento, California, 1998.

Department of Water Resources, *Flood Warnings: Responding to California's Flood Crisis,* Sacramento, California, January 2005a.

Department of Water Resources, "DWR Releases Draft EIS/EIR for South Delta Improvements Program," news release, November 10, 2005b.

Department of Water Resources, *California Water Plan Update,* Bulletin 160-05, Sacramento, California, December 2005c.

Department of Water Resources, "Progress on Incorporating Climate Change into Management of California's Water Resources," DWR Technical Memorandum Report, Sacramento, California, 2006.

Department of Water Resources et al., "Demonstration of Techniques for Reversing the Effects of Subsidence in the Sacramento–San Joaquin Delta," CALFED Ecosystem Restoration Program (ERP- 98-C01), draft annual report, 2002.

Dettinger, M. D., "From Climate-Change Spaghetti to Climate-Change Distributions for 21st Century California," *San Francisco Estuary and Watershed Science,* Vol. 3, Article 4, 2005, available at http://repositories. cdlib.org/jmie/sfews/vol3/iss1/art4.

Dettinger, M. D., D. R. Cayan, M. K. Meyer, and A. E. Jeton, "Simulated Hydrologic Responses to Climate Variations and Change in the Merced, Carson and American River Basins, Sierra Nevada, California, 1900–2099," *Climatic Change,* Vol. 62, 2004, pp. 283–317.

Deverel, S. J., and S. Rojstaczer, "Subsidence of Agricultural Lands in the Sacramento–San Joaquin Delta, California: Role of Aqueous and Gaseous Carbon Fluxes," *Water Resources Research,* Vol. 32, 1996, pp. 2359–2367.

Deverel, S. J., B. Wang,, and S. A. Rojstaczer, "Subsidence of Organic Soils, Sacramento–San Joaquin Delta," in J. W. Borchers, ed., *Land Subsidence Case Studies and Current Research. Proceedings of the Joseph Poland Subsidence Symposium,* Association of Engineering Geologists, Sudbury, Massachusetts, 1998, pp. 489–502.

Dileanis, P. D., K. P. Bennett, and J. L. Domagalski, "Occurrence and Transport of Diazinon in the Sacramento River, California, and Selected Tributaries During Three Winter Storms, January–February 2000," U.S. Geological Survey Report WRIR - 02-4101, Sacramento, California, 2002, available at http://water.usgs.gov/pubs/wri/wri02-4101.

Draper, A. J., M. W. Jenkins, K. W. Kirby, J. R. Lund, and R. E. Howitt, "Economic-Engineering Optimization for California Water Management," *Journal of Water Resources Planning and Management,* Vol. 129, No. 3, May 2003, pp. 155–164.

East Bay Municipal Utilities District, *Urban Water Management Plan,* Oakland, California, 2005.

Flyvbjerg, Bent, Nils Bruzelius, and Werner Rothengatter, *Megaprojects and Risk: An Anatomy of Ambition,* Cambridge University Press, Cambridge, United Kingdom, 2003.

Fox, J. P., T. R. Mongan, and W. J. Miller, "Trends in Freshwater Inflow to San Francisco Bay from the Sacramento–San Joaquin Delta," *Water Resources Bulletin,* Vol. 26, No. 1, 1990, pp. 101–116.

Gardner, Michael, "Would Canal Ease the Risk to Our Water Supply? A Proposed $3 Billion Bond Measure Tied to State Levee Fears Revives Old Controversies, Including a Plan for a New North-South Delivery System," *Torrance Daily Breeze,* April 17, 2006.

Grunwald, Michael, *The Swamp: The Everglades, Florida, and the Politics of Paradise,* Simon and Schuster, New York, 2006.

Hanak, Ellen, *Who Should Be Allowed to Sell Water in California? Third Party Issues and the Water Market,* Public Policy Institute of California, San Francisco, California, 2003.

Hayhoe, K., et al., "Emissions Pathways, Climate Change, and Impacts on California," *Proceedings of the National Academy of Sciences,* Vol. 101, 2004, pp. 12422–12427.

Herbold, B., A. D. Jassby, and P. B. Moyle, *Status and Trends Report on Aquatic Resources in the San Francisco Estuary,* San Francisco Estuary Project, Oakland, California, 1992.

Herbold, B., and P. B. Moyle, *Ecology of the Sacramento–San Joaquin Delta: A Community Profile,* U.S. Fish and Wildlife Service Biological Report, Vol. 85, No. 7.22, September 1989.

Hoge, Patrick, "Environmental Group Sues to Block Below-Sea-Level Housing Tract," *San Francisco Chronicle,* April 12, 2006a.

Hoge, Patrick, "Suit Filed over Plan to Build Homes in Flood Zone; Environmental Groups Say Lathrop Project Was OKd Illegally, Would Put Lives at Risk," *San Francisco Chronicle,* May 23, 2006b.

Holling, C. S., ed., *Adaptive Environmental Assessment and Management,* John Wiley & Sons, New York, 1978.

Hooke, R. L., "On the History of Humans as Geomorphic Agents," *Geology,* Vol. 28, 2000, pp. 843–846.

Howitt, R. E., "Positive Mathematical-Programming," *American Journal of Agricultural Economics,* Vol. 77, No. 2, May 1995, pp. 329–342.

Howitt, R. E., and S. Msangi, "Reconstructing Disaggregate Production Functions," AAEA-WAEA Annual Meeting, Long Beach, California, 2002.

Howitt, R. E., K. B. Ward, and S. M. Msangi, "Statewide Water and Agricultural Production Model." Department of Agricultural and Resource Economics, University of California, Davis, 2001.

Howitt, R. E., M. Tauber, and E. Pienaar, *Impacts of Global Climate Change on California's Agricultural Water Demand,* Department of Agricultural and Resource Economics, University of California, Davis, May 8, 2003.

Hundley, Norris, *The Great Thirst: Californians and Water, A History,* 2nd ed., University of California Press, Berkeley, California, 2001.

Illingworth, W., R. Mann, and S. Hatchet, "Economic Consequences of Water Supply Export Disruption Due to Seismically Initiated Levee Failures in the Delta," Appendix B in Jack R. Benjamin & Associates, *Preliminary Seismic Risk Analysis Associated with Levee Failures in the Sacramento–San Joaquin Delta,* Menlo Park, California, 2005.

Intergovernmental Panel on Climate Change, "Climate Change 2001: The Scientific Basis," in *IPCC Third Assessment Report: Climate Change 2001,* Cambridge University Press, Cambridge, United Kingdom, 2001.

Jack R. Benjamin and Associates, *Preliminary Seismic Risk Analysis Associated with Levee Failures in the Sacramento–San Joaquin Delta,* report prepared for the California Bay-Delta Authority and the California Department of Water Resources, Menlo Park, California, June 2005.

Jackson, W. T., and A. M. Paterson, *The Sacramento–San Joaquin Delta and the Evolution and Implementation of Water Policy: An Historical Perspective,* California Water Resources Center, Contribution No. 163, University of California, Davis, June 1977.

Jain, S., M. Hoerling, and J. Eischeid, "Decreasing Reliability and Increasing Synchroneity of Western North American Streamflow," *Journal of Climate,* Vol. 18, 2005, pp. 613–618.

Jenkins, M. W., A. J. Draper, J. R. Lund, R. E. Howitt, S. K. Tanaka, R. Ritzema, G. F. Marques, S. M. Msangi, B. D. Newlin, B. J. Van Lienden, M. D. Davis, and K. B. Ward, *Improving California Water Management: Optimizing Value and Flexibility,* Center for Environmental and Water Resources Engineering Report No. 01-1, Department of Civil and Environmental Engineering, University of California, Davis, 2001, available at http://cee.engr.ucdavis.edu/faculty/lund/CALVIN/.

Jenkins, M. W., J. R. Lund, and R. E. Howitt, "Economic Losses for Urban Water Scarcity in California," *Journal of the American Water Works Association,* Vol. 95, No. 2, February 2003, pp. 58–70.

Jenkins, M. W., J. R. Lund, R. E. Howitt, A. J. Draper, S. M. Msangi, S. K. Tanaka, R. S. Ritzema, and G. F. Marques, "Optimization of California's Water System: Results and Insights," *Journal of Water Resources Planning and Management,* Vol. 130, No. 4, July 2004, pp. 271–280.

Johnson, Hans, "California's Population in 2025," in Ellen Hanak and Mark Baldassare, eds., *California 2025: Taking on the Future,* Public Policy Institute of California, San Francisco, California, 2005.

Kelley, Robert, *Battling the Inland Sea,* University of California Press, Berkeley, California, 1989.

Kimmerer, W., "Open Water Processes of the San Francisco Estuary: From Physical Forcing to Biological Responses," *San Francisco Estuary and Watershed Science,* Vol. 2, No. 1, 2004, available at http://repositories.cdlib.org/jmie/sfews/vol2/iss1/art1.

Knowles, N., "Natural and Management Influences on Freshwater Inflows and Salinity in the San Francisco Estuary at Monthly to Interannual Scales," *Water Resources Research,* Vol. 38, No. 12, 2002.

Knowles, N., and D. Cayan, "Elevational Dependence of Projected Hydrologic Changes in the San Francisco Estuary and Watershed," *Climatic Change,* Vol. 62, 2004, pp. 319–336.

Landis, J. D., and M. Reilly, "How We Will Grow: Baseline Projections of California's Urban Footprint Through the Year 2100," Project Completion Report, Department of City and Regional Planning, Institute of Urban and Regional Development, University of California, Berkeley, 2002.

Leavenworth, Stuart, "Flood Risk Is Seen: Army Corps Official Says Unwise Development May Cost Taxpayers Dearly," *Sacramento Bee,* June 16, 2004a.

Leavenworth, Stuart, "High Hopes in the Delta: Developer Insists Super Levees Can Protect Homes on Flood-Prone Island," *Sacramento Bee,* July 4, 2004b.

Leavenworth, Stuart, "Dire Warnings for Delta's Future: UCD Scientists Say the State Needs to Prepare for Likely Levee Failures," *Sacramento Bee,* October 5, 2004c.

Light, T., E. D. Grosholz, and P. B. Moyle, *Delta Ecological Survey (Phase I): Nonindigenous Aquatic Species in the Sacramento–San Joaquin Delta, a Literature Review,* Final report for agreement DCN 1113322J011, U.S. Fish and Wildlife Service, Stockton, California, 2005.

Little Hoover Commission, *Still Imperiled, Still Important. The Little Hoover Commission's Review of the CALFED Bay-Delta Program,* Sacramento, California, November 2005.

Logan, S. H., "Simulating Costs of Flooding Under Alternative Policies for the Sacramento–San Joaquin Delta," *Water Resources Research,* Vol. 26, 1990, pp. 799–809.

Lopez, C. B., J. E. Cloern, T. A. Schraga, A. J. Little, L. V. Lucas, J. K. Thompson, and J. R. Burau, "Ecological Values of Shallow-Water Habitats: Implications for the Restoration of Disturbed Ecosystems," *Ecosystems,* Vol. 9, 2006, pp. 422–444.

Lowe, S., M. Browne, and S. Boudjelas, "100 of the World's Worst Invasive Alien Species: A Selection from the Global Invasive Species Database," IUCN, Invasive Species Specialist Group, Auckland, New Zealand, 2005.

Lucas, Greg, "Katrina Levee Breach Dredges Up Canal Debate; Peripheral Project Rejected in 1982—New Proposals Met with Tepid Response," *San Francisco Chronicle,* November 14, 2005.

Lucas, L. V., J. E. Cloern, J. K. Thompson, and N. E. Monsen, "Functional Variability of Shallow Tidal Habitats in the Sacramento–San Joaquin Delta: Restoration Implications," *Ecological Applications,* Vol. 12, 2002, pp. 1528–1547.

Lund, J. R., R. E. Howitt, M. W. Jenkins, T. Zhu, S. K. Tanaka, M. Pulido, M. Tauber, R. Ritzema, and I. Ferreira, "Climate Warming and California's Water Future," Center for Environmental and Water Resources Engineering Report No. 03-1, Department of Civil and Environmental Engineering, University of California, Davis, 2003, available at http://cee.engr.ucdavis.edu/faculty/lund/CALVIN/.

Machado, Michael, "Letter to Editor: Water Plans Need Results," *San Joaquin Record,* August 27, 2003.

Machado, Michelle, "Panelists Agree: No More Building Around the Delta. Participants Represented North, South Water Concerns," *Stockton Record,* October 9, 2005.

Matern, S. A., P. B. Moyle, and L. C. Pierce, "Native and Alien Fishes in a California Estuarine Marsh: Twenty-One Years of Changing Assemblages," *Transactions of the American Fisheries Society,* Vol. 131, 2002, pp. 797–816.

Matthew, Raymond, *Variation and Control of Salinity in Sacramento–San Joaquin Delta and Upper San Francisco Bay,* Bulletin 27, Division of Water Resources, California Department of Public Works, Sacramento, California, 1931a.

Matthew, Raymond, *Economic Aspects of a Salt Water Barrier Below the Confluence of Sacramento and San Joaquin Rivers,* Bulletin 28, Division of Water Resources, California Department of Public Works, Sacramento, California, 1931b.

Maurer, E. P., "Uncertainty in Hydrologic Impacts of Climate Change in the Sierra Nevada, California Under Two Emissions Scenarios," *Climatic Change,* 2006 (in press).

Medellin, J., J. Harou, M. Olivares, J. R. Lund, R. Howitt, S. Tanaka, M. Jenkins, K. Madani, and T. Zhu, *Climate Warming and Water Supply Management in California,* White Paper CEC-500-2005-195-SD,

Climate Change Program, California Energy Commission, Sacramento, California, February 2006.

Meral, Gerald H., "The Delta Facility," unpublished document, Inverness, California, August 19, 2005a.

Meral, Gerald H., "Guest Opinion: We Must Save the Delta—But How? Diverting Water Through a Peripheral Pipeline Would Benefit Ecosystem, Water Users," *Sacramento Bee,* October 16, 2005b.

Metropolitan Water District of Southern California, "Long-Term Sustainability in the Sacramento–San Joaquin River Delta. Policy Principles of the Metropolitan Water District of Southern California," Board Meeting document, Los Angeles, California, April 11, 2006.

Miller, N. L., K. E. Bashford, and E. Strem, "Potential Impacts of Climate Change on California Hydrology," *Journal of the American Water Resources Association,* Vol. 39, No. 4, 2003, pp. 771–784.

Montgomery, Ben, "Guest Opinion: The Wisdom of Delta Development; PRO: At Home Behind Super Levees," *San Francisco Chronicle,* October 8, 2006.

Morton, B., and K. Y. Tong, "The Salinity Tolerance of *Corbicula fluminea* (Bivalva: Corbiculoidea) from Hong Kong," *Malocological Review,* Vol. 18, 1985, pp. 91–95.

Mount, J. F., and R. Twiss, "Subsidence, Sea Level Rise, Seismicity in the Sacramento–San Joaquin Delta," *San Francisco Estuary and Watershed Science,* Vol. 3, Article 5, 2005, available at http://repositories.cdlib.org/jmie/sfews/vol3/iss1/art5.

Mount, J. F., R. Twiss, and R. A. Adams, "The Role of Science in the Delta Visioning Process," draft report of the Delta Science Panel of the CALFED Bay-Delta Science Program, 2006, available at http://science.calwater.ca.gov/pdf/CSP_delta_vision_process_Twiss_062306.pdf.

Moyle, P. B., *Inland Fishes of California. Revised and Expanded,* University of California Press, Berkeley, Berkeley, California, 2002.

Moyle, P. B., and M. P. Marchetti, "Predicting Invasion Success: Freshwater Fishes in California as a Model," *Bioscience*, Vol. 56, 2006, pp. 515–524.

Moyle, P. B., R. D. Baxter, T. Sommer, T. C. Foin, and S. A. Matern, "Biology and Population Dynamics of Sacramento Splittail (*Pogonichthys macrolepidotus*) in the San Francisco Estuary: A Review," *San Francisco Estuary and Watershed Science*, Vol. 2, No. 2, 2004, pp. 1–47, available at http://repositories.cdlib.org/jmie/sfews/vol2/iss2/art3.

Moyle P. B., P. K. Crain, and K. Whitener, "Patterns in the Use of a Restored California Floodplain by Native and Alien Fishes," *San Francisco Estuary and Watershed Science,* in press, available at http://repositories.cdlib.org/jmie/sfews/.

Nelson, Barry, "Guest Opinion: Building a Canal Would Eliminate Southern California's Major Motive to Protect the Delta," *Sacramento Bee,* October 16, 2005.

Newlin, B. D., M. W. Jenkins, J. R. Lund, and R. E. Howitt, "Southern California Water Markets: Potential and Limitations," *Journal of Water Resources Planning and Management,* Vol. 128, No. 1, January/February 2002, pp. 21–32.

Nobriga, M. L., F. Feyrer, R. D. Baxter, and M. Chothowski, "Fish Community Ecology in an Altered River Delta: Spatial Patterns in Species Composition, Life History Strategies, and Biomass," *Estuaries*, Vol. 28, 2005, pp. 776–785.

Null, S., and J. R. Lund, "Re-Assembling Hetch Hetchy: Water Supply Implications of Removing O'Shaughnessy Dam," *Journal of the American Water Resources Association,* Vol. 42, No. 4, April 2006, pp. 395–408.

Orlob, G. T., "An Alternative to the Peripheral Canal," *Journal of the Water Resources Planning and Management Division,* ASCE, Vol. 108, No. WR1, March 1982, pp. 123–141.

Orr M., S. Crooks, and P. B. Williams, "Will Restored Tidal Marshes Be Sustainable?" in L. R. Brown, ed., *Issues in San Francisco Estuary Tidal Wetlands Restoration, San Francisco Estuary and Watershed Science,* Vol. 1, Article 5, 2003, available at http://repositories.cdlib.org/jmie/sfews/vol1/iss1/art5/.

Parchaso, F., and J. K. Thompson, "Influence of Hydrologic Processes on Reproduction of the Introduced Bivalve *Potamocorbula amurensis* in Northern San Francisco Bay, California," *Pacific Science*, Vol. 56, 2002, pp. 329–345.

Pethick, J. S., and S. Crook, "Development of a Coastal Vulnerability Index: A Geomorphological Perspective," *Environmental Conservation,* Vol. 27, 2000, pp. 359–367.

Pitzer, Gary, "Developing a Delta Vision," *Western Water,* May/June 2006.

Plater, J., and W. W. Wade, "Estimating Potential Demand for Freshwater Recreation Activities in the Sacramento–San Joaquin Delta 1997–2020," *Energy and Water Economics,* Columbia, Tennessee, 2002, available at www.dbw.ca.gov.

Pollard, Vic, "Valley Boost on Tap. Plan Between State Projects to Increase Water for Kern as Much as 50,000 Acre-Feet a Year," *Bakersfield Californian,* August 20, 2003.

Port of Sacramento, *Overview,* Sacramento, California, 2006, available at www.portofsacramento.com.

Port of Stockton, *Annual Report for 2004,* Stockton, California, 2004, available at www.portofstockton.com.

Pulido-Velázquez, M., M. W. Jenkins, and J. R. Lund, "Economic Values for Conjunctive Use and Water Banking in Southern California," *Water Resources Research,* Vol. 40, No. 3, March 2004.

Ramsey, F., "A Contribution to the Theory of Taxation," *Economic Journal,* Vol. 37, March 1927, pp. 47–61.

Reed, D. J., "Sea-Level Rise and Coastal Marsh Sustainability: Geological and Ecological Factors in the Mississippi Delta Plain," *Geomorphology,* Vol. 48, 2002, pp. 233–243.

Rosekrans, Spreck, and Ann H. Hayden, *Finding the Water: New Water Supply Opportunities to Revive the San Francisco Bay–Delta Ecosystem,* Environmental Defense, Oakland, California, 2005.

Rosenzweig, M., *Win-Win Ecology: How the Earth's Species Can Survive in the Midst of Human Enterprise,* Oxford University Press, Oxford, United Kingdom, 2004.

Salgado-Maldonado, Guillermo, and Raúl F. Pineda-López, "The Asian Fish Tapeworm *Bothriocephalus acheilognathi*: A Potential Threat to Native Freshwater Fish Species in Mexico," *Biological Invasions,* Vol. 5, 2003, pp. 261–268.

San Francisco Public Utilities Commission, "Urban Water Management Plan," San Francisco, California, 2005, available at www.sfwater.org.

Santa Clara Valley Water District, "Urban Water Management Plan," San Jose, California, 2005, available at www.valleywater.org.

Schofield, P. J., J. D. Williams, L. G. Nico, P. Fuller, and M. R. Thomas, *Foreign Nonindiginous Carps and Minnows (Cyprinidae) in the United States—A Guide to Their Identification, Distribution, and Biology,* USDI, USGS Special Investigations Report 2005-5041, 2005.

Seneca, R. S., "Inherent Advantage, Costs and Resource Allocation in the Transportation Industry," *American Economic Review,* Vol. 63, No. 5, December 1973, pp. 945–956.

Shlemon R. J., and E. L. Begg, "Late Quaternary Evolution of the Sacramento–San Joaquin Delta, California," in R. P. Suggate, and M. M. Cressel, eds., *Quaternary Studies,* Bulletin 13, The Royal Society of New Zealand, 1975, pp. 259–266.

Snow, Lester, "Protecting Sacramento/San Joaquin Bay–Delta Water Supplies and Responding to Failures in California Water Deliveries," testimony before the U.S. House of Representatives Committee on Resources, Subcommittee on Water and Power, Washington, D.C., April 6, 2006, available at http://www.publicaffairs.water.ca.gov.

Sommer, T. R., W. C. Harrell, M. Nobriga, R. Brown, P. B. Moyle, W. J. Kimmerer, and L. Schemel, "California's Yolo Bypass: Evidence That Flood Control Can Be Compatible with Fish, Wetlands, Wildlife and Agriculture," *Fisheries*, Vol. 58, No. 2, 2001a, pp. 25–33.

Sommer, T. R., M. L. Nobriga, W. C. Harrell, W. Batham, and W. J. Kimmerer, "Floodplain Rearing of Juvenile Chinook Salmon: Evidence of Enhanced Growth and Survival," *Canadian Journal of Fisheries and Aquatic Sciences,* Vol. 58, 2001b, pp. 325–333.

Sommer, T. R., W. C. Harrell, A. Mueller-Solger, B. Tom, and W. Kimmerer, "Effects of Flow Variation on Channel and Floodplain Biota and Habitats of the Sacramento River, California, USA," *Aquatic Conservation: Marine and Freshwater Ecosystems,* Vol. 14, 2004, pp. 247–261.

State Water Commission, *Report*, Sacramento, California, 1917.

State Water Resources Control Board, Order WR 2006-0006, Sacramento, California, February 15, 2006.

Stewart, I., D. R. Cayan, and M. D. Dettinger, "Changes in Snowmelt Runoff Timing in Western North America under a 'Business as Usual' Climate Change Scenario," *Climate Change,* Vol. 62, 2004, pp. 217–232.

Stockton Record, "Editorial: This Plan Should Stay on Delta's Periphery," July 11, 2005.

Stockton Record, "Editorial: Plan Doesn't Hold Water," April 27, 2006.

Tanaka, S. K., and J. R. Lund, "Effects of Increased Delta Exports on Sacramento Valley's Economy and Water Management," *Journal of the American Water Resources Association,* Vol. 39, No. 6, December 2003, pp. 1509–1519.

Tanaka, S. K., T. Zhu, J. R. Lund, R. E. Howitt, M. W. Jenkins, M. A. Pulido, M. Tauber, R. S. Ritzema, and I. C. Ferreira, "Climate Warming and Water Management Adaptation for California," *Climatic Change,* Vol. 76, No. 3-4, June 2006.

Taugher, Mike, "Lawmaker Calls $8 Billion Water Plan Flawed," *Contra Costa Times,* December 9, 2004.

Taugher, Mike, "Delta in Decline: Part 5; A Struggle to Quench State's Thirst for Water," *Contra Costa Times,* December 29, 2005.

Taugher, Mike, "Pump Proposal Threatens Water Pact," *Contra Costa Times,* January 25, 2006a.

Taugher, Mike, "Peripheral Canal Plan Resurfaces," *Contra Costa Times,* April 13, 2006b.

Thompson, Don, "Saving Tiny Delta Fish Carries Heavy Price Tag, Report Says," *Sacramento Bee,* October 20, 2005a.

Thompson, Don, "Delta Earthquake Could Halt Flow of Water in State; Levees Vulnerable, Top Official Warns," *San Diego Union Tribune,* November 2, 2005b.

Thompson, John, *Settlement Geography of the Sacramento–San Joaquin Delta,* California, dissertation, Stanford University, 1957.

Torres, R. A., et al., *Seismic Vulnerability of the Sacramento–San Joaquin Delta Levees,* report of Levees and Channels Technical Team, Seismic Vulnerability Sub-Team, to CALFED Bay-Delta Program, Sacramento, California, 2000.

U.S. Bureau of Reclamation, "Central Valley Project Improvement Act Programmatic Environmental Impact Statement," CD-ROM, Sacramento, California, 1997.

U.S. Bureau of Reclamation, "CALSIM II San Joaquin River Model (DRAFT-online version)," U.S. Department of the Interior, Mid Pacific

Region, Sacramento, California, April, 2005, available at http://science. calwater.ca.gov/pdf/calsim/CALSIMSJR_DRAFT_072205.pdf.

Van Der Most, Herman, and Mark Wehrung, "Dealing with Uncertainty in Flood Risk Assessment of Dike Rings in the Netherlands," *Natural Hazards,* Vol. 36, 2005, pp. 191–206.

van Genuchten, M. Th., and G. J. Hoffman, "Analysis of Crop Salt Tolerance Data," in I. Shainberg and J. Shalbert, eds., *Soil Salinity under Irrigation, Processes and Management,* Ecological Studies 51, Springer Verlag, New York, 1984, pp. 258–271.

Van Lienden, B., and J. R. Lund, "Spatial Complexity and Reservoir Optimization Results," *Civil and Environmental Engineering Systems,* Vol. 21, No. 1, March 2004, pp. 1–17.

VanRheenan, N. T., A. W. Wood, R. N. Palmer, and D. P. Lettenmaier, "Potential Implications of PCM Climate Change Scenarios for Sacramento–San Joaquin River Basin Hydrology and Water Resources," *Climate Change,* Vol. 62, 2004, pp. 257–281.

Weiser, Matt, "Delta Danger; A Decline in Fish Species and Their Food Source Is a Reminder of a Recurring Worry in the West: A Broad Ecosystem Collapse," *Sacramento Bee,* July 3, 2005.

Weiser, Matt, "New Push to Save Delta Smelt. Three Groups Petition for Fish to Receive 'Endangered' Status," *Sacramento Bee,* March 9, 2006a.

Weiser, Matt, "DWR Sued by Fishing Alliance; Fish: DWR Working on New Plan, *Sacramento Bee,* October 5, 2006b.

Weiser, Matt, "Sugar Mill Project Halted; Two Groups Appeal Yolo's Approval of Clarksburg Tract, Citing Delta Risks," *Sacramento Bee,* November 8, 2006c.

Werner, L., "The Influence of Salinity on the Heat-Shock Protein Response of *Potamocorbula amurensis* (Bivalva)," *Marine Environmental Research,* Vol. 58, 2004, pp. 803-807.

Wright, Patrick, "Letters to the Editor: Bold Water Plan," *Sacramento Bee,* February 4, 2004.

Young, Samantha, "Delta Water Supply May Be Cut; Environmentalists Act to Protect Tiny Fish's Habitat," *Associated Press,* reprinted in *Los Angeles Daily News,* July 8, 2006.

Young, Walker R., *Report on Salt Water Barrier Below Confluence of Sacramento and San Joaquin Rivers, California,* Bulletin 22, Vols. I and II, Division of Water Resources, California Department of Public Works, Sacramento, California, 1929.

Zhu, T., M. W. Jenkins, and J. R. Lund, "Estimated Impacts of Climate Warming on California Water Availability," *Journal of the American Water Resources Association,* Vol. 41, No. 5, October 2005, pp. 1027–1038.

About the Authors

WILLIAM E. FLEENOR

William E. Fleenor is a professional research engineer in the Civil and Environmental Engineering Department at the University of California, Davis. He received a bachelor's degree in mechanical engineering from the Rose-Hulman Institute of Technology. After a career in engineering sales and marketing, he earned a master's degree in environmental engineering from UC Davis and a Ph.D. in water resources. He has been involved with various hydrodynamic and water quality research projects involving the Delta and is currently the project manager for two CALFED Bay-Delta-funded water quality modeling efforts.

ELLEN HANAK

Ellen Hanak is a research fellow and director of the Economy Program at the Public Policy Institute of California. Her career has focused on the economics of natural resource management and agricultural development. At PPIC, she has launched a research program on water policy and has published reports and articles on water marketing, water and land use planning, and water conservation. Before joining PPIC in 2001, she held positions at the Center for Cooperation in International Agricultural Development in France, on the President's Council of Economic Advisers, and at the World Bank. She holds a Ph.D. in economics from the University of Maryland.

RICHARD E. HOWITT

Richard E. Howitt is professor and department chair of Agricultural and Resource Economics at the University of California, Davis. He teaches both graduate and undergraduate courses in resource economics, economic theory, and operations research. His current research interests include constructing disaggregated economic modeling methods based on maximum entropy estimators, testing the allocation of water resources by market mechanisms, and developing empirical dynamic stochastic methods to analyze changes in investments and institutions. He serves on advisory boards for the California Department of Water Resources and the U.S. Academy of Sciences.

JAY R. LUND

Jay R. Lund is a professor in the Civil and Environmental Engineering Department at the University of California, Davis. He specializes in the management of water and environmental systems. His activities have included system optimization studies for California, the Columbia River, the Missouri River, and several other systems—as well as studies of climate change adaptation, water marketing, water conservation, water utility planning, and reservoir operations. He was on the Advisory Committee for the 1998 and 2005 California Water Plan Updates, is a former editor of the *Journal of Water Resources Planning and Management*, and has authored or co-authored over 200 publications.

JEFFREY F. MOUNT

Jeffrey F. Mount is a professor in the Geology Department at the University of California, Davis, where he has worked since 1980. His research and teaching interests include fluvial geomorphology, conservation and restoration of large river systems, flood plain management, and flood policy. He holds the Roy Shlemon Chair in Applied Geosciences at UC Davis, is the director of the UC Davis Center for Watershed Sciences, and chairs the CALFED Independent Science Board. He is author of *California Rivers and Streams: The Conflict between Fluvial Process and Land Use* (UC Press, 1995).

PETER B. MOYLE

Peter B. Moyle has been studying the ecology and conservation of freshwater and estuarine fish in California since 1969 and has focused on the San Francisco Estuary since 1976. He was head of the Delta Native Fishes Recovery Team and a member of the Science Board for the CALFED Ecosystem Restoration Program. He has authored or coauthored over 160 scientific papers and five books, including *Inland Fishes of California* (UC Press, 2002). He is a professor of fish biology in the Department of Wildlife, Fish, and Conservation Biology at the University of California, Davis, and is associate director of the UC Davis Center for Watershed Sciences.

Related PPIC Publications

Water for Growth: California's New Frontier (2005)
Ellen Hanak

California 2025: Taking on the Future (2005)
Ellen Hanak, Mark Baldassare (editors)

Who Should Be Allowed to Sell Water in California? Third-Party Issues and the Water Market (2003)
Ellen Hanak

Making Room for the Future: Rebuilding California's Infrastructure (2003)
David E. Dowall, Jan Whittington

California's Infrastructure Policy for the 21st Century: Issues and Opportunities (2000)
David E. Dowall

PPIC publications may be ordered by phone or from our website
(800) 232-5343 (mainland U.S.)
(415) 291-4400 (outside mainland U.S.)
www.ppic.org